Physiological Disorders of Horticultural Crops

NIPA GENX ELECTRONIC RESOURCES & SOLUTIONS P. LTD.
New Delhi-110 034

About the Authors

Dr. Pankaj Kumar Ray is Subject Matter Specialist (Horticulture) at Krishi Vigyan Kendra, Saharsa of Bihar Agricultural University, Sabour (Bhagalpur), India. In 2012 he received Master's Degree in Vegetable Science from Acharya Narendra Deva University of Agriculture and Technology, Kumarganj, Ayodhya (India) and in 2017 he received Doctorate in Horticulture from Bihar Agricultural University, Sabour, Bhagalpur, Bihar (India). Dr. Ray has supervised many research projects on horticulture crops as P.I./ Co-P.I. at B.A.U., Sabour with the goal of INM in fruit crops, Makhana, Vegetable crops as well as engaged in Climate resilient variety & technology demonstration at farmers' field. He is accompanied with development of two brinjal varieties and one garlic variety, apart from integrated weed management technology in onions. Based on his extension performance at farmers' field, he was awarded the Best Extension Scientist Award (2020) by BAU, Sabour beside this, he is also credited with 09 International/National/ Institutional Awards. Dr. Ray has successfully organised many training programmes and conferences. He has published 26 research papers, 3 review papers, 3 souvenir papers, 35 book chapters, and 25 popular articles. He is also a life member of several reputed societies and editorial boards member/ reviewer for many prestigious journals.

Dr. Shashank Shekhar Solankey is presently working as Assistant Professor–cum–Jr. Scientist (Horticulture: Vegetable Science) at Dr. Kalam Agricultural College, Kishanganj of Bihar Agricultural University, Sabour (Bhagalpur), India. He is also Nodal officer of newly established Horticulture Research Centre, Dr. Kalam Agricultural College, Kishanganj, Bihar. He completed his Master's Degree in Vegetable Science from Acharya Narendra Deva University of Agriculture and Technology, Kumarganj, Ayodhya (India) in 2006 and Doctorate in Horticulture from Banaras Hindu University, Varanasi (India) in 2010. Dr. Solankey has worked as SRF as well as RA at ICAR-Indian Institute of Vegetable Research (IIVR), Varanasi (U.P.) from 2010 - 2012. He joined Bihar Agricultural University, Sabour on 17th September, 2012. He has been involved in teaching, research, extension and training activities at the university. Dr. Solankey had been assigned for development of a new college at Kishanganj, Bihar and deputed on 7th September, 2015 at newly established Dr. Kalam Agricultural College, Kishanganj under the umbrella of BAU, Sabour. He has been acting as Nodal Officer of newly established Horticulture Research Centre, Kishanganj, Bihar from 27th May, 2021 till date. He has handled four research projects on vegetable crops as P.I./ Co-P.I. at B.A.U., Sabour with the objective of biotic and abiotic stress management as well as quality improvement in solanaceous vegetables and okra. He is associated with development of two brinjal varieties (Sabour Sadabahar & Sabour Krishnakali) and one technology on management of sucking pests of okra. Dr. Solankey has supervised 4 M.Sc. students and also acted as member of advisory committee of 7 M.Sc. and 3 Ph.D. students. He has published 56 research papers, 07 review papers, 01 souvenir paper, 08 edited books, 50 book chapters, 25 popular articles. He is also life time member of Horticulture Society of India, New Delhi; Indian Society of Vegetable Science, IIVR, Varanasi; International Society for Noni Science, Perungudi, Chennai; Society for Scientific Development in Agriculture & Technology, Meerut and Bihar Horticulture Society, BAU, Bihar. He is also reviewer of International Journal of Plant & Soil Science as well as Scientia Horticulturae. Dr. Solankey has also been awarded with Best Teacher Award (2016) as well as Best Researcher Award (2016) by Bihar Agricultural University, Sabour beside these he has 12 other awards/ recognitions also.

Physiological Disorders of Horticultural Crops Causes and Management

P.K. Ray
Subject Matter Specialist (Horticulture)
Krishi Vigyan Kendra, Saharsa
Bihar Agricultural University
Sabour, Bhagalpur

S.S. Solankey
Assistant Professor
(Horticulture: Vegetable Science)
Dr. Kalam Agricultural College, Kishanganj, Bihar
Bihar Agricultural University
Sabour (Bhagalpur), Bihar

NIPA GENX ELECTRONIC RESOURCES & SOLUTIONS P. LTD.
New Delhi-110 034

NIPA GENX ELECTRONIC RESOURCES & SOLUTIONS P. LTD.

101,103, Vikas Surya Plaza, CU Block
L.S.C.Market, Pitam Pura, New Delhi-110 034
Ph : +91 11 27341616, 27341717, 27341718
E-mail:newindiapublishingagency@gmail.com
www: www.nipabooks.com

For customer assistance, please contact
Phone: + 91-11-27 34 17 17 Fax: + 91-11- 27 34 16 16
E-Mail: feedbacks@nipabooks.com

ISBN: 978-93-91383-12-1

Composed and Designed by NIPA.

Preface

Horticulture harvests are an important part of our diet that presently emerged as growth engine of the Indian as well as global economy. In India horticulture sector pays 38 per cent of the agricultural GDP and supports nearly 20 per cent of the agricultural work force with less than 14 per cent of the total cropped area. Horticulture crops such as vegetables, fruits, flowers, spices, condiments and plantation crops are an integral part of food, nutritional, and economic securities in India. The area under horticultural crops grew by 2.6 per cent per annum and annual production increased by 4.8 per cent in the last decade. The production of horticultural crops was at a high record of 320.48 MT with 25.66 million hectares area during the year 2019-20. India is the second largest producer of fruits and vegetables that account for nearly 90 per cent of total horticultural production with significant recognition in international market. The development achieved in the horticulture sector indicates the fact that there is a growing demand for horticulture crops.

Nevertheless, there are many limitations of horticultural production that is a serious barrier in obtaining optimum yield in which physiological disorders seriously affects the quality as well as quantity of the horticultural produce. Moreover, a very limited information is available on the physiological disorders, causes and management of horticultural crops. In fact, the physiological or abiotic disorders are principally caused by changing climatic conditions such as extreme heat and cold, flood, drought, disturbed soil nutrients including soil pH, harmful gasses, etc. In the present era of global warming, physiological disorder refers to a wide array of plant problems that are not caused by the biological agents such as insects or pathogens but rather by climatic factors (temperature, rain, humidity, light, etc.) and management practices (training, pruning, irrigation, fertilization, harvest procedure, postharvest management, etc.) that change the microclimate endured by the plant. These defects are the result of plant responses to abnormal environmental conditions occurred during the pre and/ or post-harvest seasons that lead to undesirable changes in the cell physiology and fruit quality and are also known as abiotic disorders. These abiotic disorders are hard to diagnose because of the combined and subtle nature of many physiological disorders in horticultural crops. These physiological disorders are the result of adverse environmental factors during field or plant growth or during processing, storage, transportation and marketing, resulting in higher economic losses.

Hence, the present book entitled "Physiological Disorders of Horticultural Crops: *Causes and Management*" provide a comprehensive knowledge of cause, symptoms and management of different physiological disorders in fruits, vegetables, flowers, spices, condiments and plantation crops. The book contains adequate symptoms along with appropriate photographs about the physiological disorders of important horticultural crops and their adequate management practices. This book addresses the current issues that are divided in six sections:

Section A: Introduction to physiological disorders and its impact on horticultural crops

Section B: Major physiological disorders of vegetable crops

Section C: Major physiological disorders of fruit crops

Section D: Major physiological disorders of flowers

Section E: Major physiological disorders of plantation crops

Section F: Major physiological disorders of spices and condiments

The above sections covered the comprehensive information on causes, symptoms and recent remedial measures of major physiological disorders of horticultural crops along with multiple choice questions answers and relevant appendices for better understanding.

The book will definitely serve as a useful guide for students, scientists, extension workers, farmers and others associated with the discipline of Horticulture.

We feel immense pleasure to express our heartfelt gratitude to Dr. Kirti Singh, Former-Chairman, ASRB, Former- Vice-Chancellor, A.N.D.U.A. & T., Ayodhya, C.S.K.H.P.A.U., Palampur and I.G.K.V., Raipur for his inspiring guidance, encouragement and blessings.

We are also thankful to our family members whose sustained support and encouragement led us to complete this book. We are also thankful to our teachers, seniors, colleagues and friends for their direct and indirect support and valuable suggestions during compilation of this book.

We would be happy if the readers may recommend any valuable suggestion to make this book more useful. The authors sincerely acknowledge the help and support taken from the various resource persons, e-resources, books and journals, authors and publishers whose publications have been used while preparing this document.

We are also thankful to NIPA (New India Publishing Agency), New Delhi for bringing out this manuscript in a very systematic manner. Finally, we are grateful to the *'Almighty God'*, for providing strength, inspirations, positive thoughts, insights, and way to complete this book.

Authors

Contents

Section 3: Major Physiological Disorders of Fruit Crops

Section 4: Major Physiological Disorders of Flowers

Section 5: Major Physiological Disorders of Plantation Crops

Section 6: Major Physiological Disorders of Spices and Condiments

List of Colour Plates

Section 1

Introduction to Physiological Disorders and its Impact on Horticultural Crops

1

General Introduction About Physiological Disorders in Horticultural Crops

Physiological plant disorder is caused by non-pathological conditions such as poor light, weather damage, waterlogging or lack of nutrients affect the operation of the system caused. Physiological condition selected from diseases caused by plant pathogens such as viruses or fungi to distinguish. Although the physiological disorder symptoms may occur such as disease, they can often be prevented by changes in environmental conditions. However, once the nutritional deficiencies of plants showing symptoms, is likely to return that season yield will be reduced.

The quality of vegetables and vegetable produce or contribute to other components of economically important part of any kind of abnormal physiological disorder is called. The reason is physiological disorders of the environment / biological factor. Physiological disorders not only cause damage themselves but also serve as the 'open door' entry for pathogens. The damage caused due to physiological disorders may be reversible or irreversible depending on time and extent/ intensity. Physiological disorders are caused because of:

- Micronutrient deficiencies
- Rapid variation in temperature
- Poor conditions of soil and
- Inappropriate water availability throughout farming

How to Identify

Physiological or abiotic disorders are separate from other disorders, because they are not caused by living organisms (viruses, bacteria, fungi, phytoplasma, insects, etc.), but grown from normal circumstances and causes a non-biological, non-biological of deviation caused distinguished. They are from what is normal, usually caused by physical or chemical changes in the plant by external factors. Some non-communicable diseases are easily identified, but others are very difficult or even impossible. Most of them, once they are irreversible in the identification of physiological disorder, it is important to know that help:

- Physiological disorder often due to lack or excess of what life-sustaining, or by the presence of interference caused some life.
- Physiological disorder will affect the plant at all phases of their lives.
- They occur with non-infectious pathogens, and therefore cannot be transferred.
- Plants react to the same agent vary greatly, from small reaction to the death.
- Physiological disorder treatment often means that treatment because the consequences of past events.
- Generally, have good tissue damage from a clear demarcation line.
- Physiological disorder itself is serious, but often as the "open door" pathogens to enter.

Types of Physiological Disorders

a) On the basis of causal factors

Various atmospheric and edaphic factors affect physiological functioning of fruit in different manners and resulted in specific type of symptoms. However, symptoms developed by a particular factor in various fruits follow almost similar pattern and can be distinguished from the symptoms developed by another factor. Different types of physiological disorders caused by variable atmospheric/ edaphic factors are listed below:

i. **Temperature related disorders:** Injuries of fruits and vegetables resulting from the reactions to temperatures outside the optimum range during the pre- and post-harvest period fall in this category. Various disorders of this nature are *'freezing injury'*, *'chilling injury'*, *'heat injury'*, etc.

ii. **Nutrient related disorders:** Imbalance supply of essential nutrients to the plant may also results in various types of physiological injuries in fruits and vegetables. For example, calcium and boron play very critical role in fruit development and harvest quality and deficiency of these nutrients lead to the physiological malfunctioning and deterioration of yield and quality in many fruits. 'Bitter pit' (apple), 'monkey fruit' (apple), 'hen and chicken' (grapes), etc are important physiological disorders caused by nutrient imbalances.

iii. **Atmosphere related disorders:** Fruit injury can also develop due to the macro and micro climatic conditions of the fruit. Very high concentration of CO_2 or low concentration of O_2 alter the normal metabolism of the plant/ plant part and result in physiological injury. Many fruits and vegetables are highly sensitive to ethylene and get deteriorated upon ethylene exposure. Besides, there may be presence of toxic chemical like ammonia, sulphur dioxide, methyl bromide, ozone, etc. which cause the damage of fruits and vegetables. Disorders of these kinds are named on the basis of causal factor like 'CO_2 injury', 'ethylene injury', 'ammonia injury', etc.

b) On the basis of appearance time

Physiological disorders of fruits may occur during both pre-harvest as well as post-harvest period and are able to cause a sizeable amount of food loss. Any physiological factor (s) like temperature, nutrients, etc., if present beyond the required range can alter the internal metabolic process of plant/ plant part but the symptoms may develop after a period of time when environmental conditions favour the development of symptoms. Thus, manifestation of fruit physiological disorder ensues in two phases called induction phase and development phase. Considering these facts in mind physiological disorders may be of three types:

i. **Pre-harvest physiological disorders:** These disorders develop in the field during fruit growth period because of one or more than one unfavorable physiological factor like temperature, nutrients, gases, etc. Examples of pre-harvest disorders are

ii. **Predetermined physiological disorders:** The primary causes for development of these disorders are some pre-harvest factors. However, the visual symptoms develop only during post-harvest storage period under such environmental conditions that encourage the symptom development. Examples of pre-harvest disorders are

iii. **Post-harvest physiological disorders:** Development of fruit disorders in response to imposed storage conditions are termed as post-harvest physiological disorder. However, temporal requirement and degree of disorder incidence again depend on pre-harvest and harvest factors.

Individual Fruit Disorders

In order to manage the physiological disorders of fruits and vegetables one should have a good comprehension of the cause (s) and biology behind the injury. A concise account of the symptom and their control measures of specific physiological disorders are discussed in the next sections/ chapters.

2

Impact of Physiological Disorders on Horticultural Crops

Apart from insects, pests and diseases, physiological disorders are one of the major threats to fruit industry which cause financial losses global. Over-all, the problems rise due to unfavorable environmental conditions and improper cultural practices which affect the normal plant growth and development is called as physiological disorders of plant. Ladaniya (2008) stated that the physiological disorder is that it is a function of physiological processes fruit tissue abnormality or failure due to abiotic stress, such as temperature, relative humidity, water stress, chemicals/ nutritional excess and deficiency. Quality and productivity of fruit affected by a greater extent, due to the physiological and nutritional disorders. Planting and yield potential quality of fruits and vegetables are affected by many physiological disorders like fruit drop, irregular bearing, fruit cracking etc. (Campostrini *et al.*, 2010) noted that disturbance in the plant metabolic activities isnearlyassociated to ecologicalfeatures which in certain cases, are problematic for farmers to stop. According to (Chatenet *et al.*, 2000) the physiological disorders which caused by different factors are mutual in several fruit and vegetable spp. The physiological disorders in different fruit and vegetable crops can be grouped in several categories based on causal aspects. The micronutrient related disorders arise in commercial orcharding due to inadequate practice of organic fertilizers, implementation of ultra-high-density planting systems, and use of unsuitable root stocks for example citrus granulation in mandarin when grafted on Jatti Khatti rootstock and imbalanced fertilizer application.

It is detected before in order to maintain fruit quality and good yields micronutrient deficiency symptoms of visual expression. In India, the defect zinc, manganese and B are common in orange, lime acid, banana, guava and papaya. To correct visual and hidden micronutrient deficiencies, the appropriate foliar and soil application is necessary. Description crops include physiological and nutritional disorders like bronzing (bronze or copper color), green (chlorophyll loss), a mold back (new leaves affecting growth cone collapse) and a number of technical terms.

Table 1: General recommendations to improve the nutritional disorders in horticultural crops

S.No.	Deficiency	Soil application (kg/ ha)	Foliar spray (%)
1.	Nitrogen	50 kg of urea application 50 kg of DAP Application	1% urea
2.	Phosphorus	25-50 kg of DAP Application	2% DAP
3.	Potassium	CAN 50 kg or Calcium chloride 60-80 kg application	KCl 1%
4.	Calcium	CAN 50 kg or Calcium chloride 60-80 kg application	0.5% $CaNO_3$ or Calcium ammonium nitrate or $CaCl_2$
5.	Sulphur	50 kg of Gypsum application	0.5% gypsum
6.	Magnesium	25-50 kg of magnesium sulphate application	0.5% $MgSO_4$ + 1% urea
7.	Iron	50-100 kg of ferrous sulphate application	$FeSO_4$ 1 to 2%
8.	Zinc	25 kg of zinc sulphate application	0.5% $ZnSO_4$
9.	Manganese	25 kg of manganese sulphate application	0.5% $MnSO_4$
10.	Copper	5-10 kg of copper sulphate application	0.1% $CuSO_4$
11.	Boron	10 kg of sodium tetraborate or borax application	0.05% borax

Section 2

Major Physiological Disorders of Vegetable Crops

3

Physiological Disorders of Solanceaus Vegetables

A. TOMATO

1. Blossom End Rot (BER)

Blossom end rot

Blossom end rot is a very common problem on green and ripe tomatoes. It started with light tan, water-soaked lesion, and then increased to black and tough. Although blossom blight itself caused only local damage, secondary biological invasions frequently lesions caused fruit rot complete. It often occurs in the largest in the hot, dry weather during fruit developing rapidly, often affect the first ripe fruit. A distal end leading to fruit blossom blight localized calcium deficiency because the calcium is not a highly mobile element, changes may occur in the water supply of a defect, even for a short period of time. Therefore, the possibility of the disease promoting extreme moisture from plant to reduce calcium intake other conditions, e.g., high salt, ammonium nitrogen and a high relative humidity may aggravate the problem. Fast-growing plants more susceptible to disease.

Management

- Single foliar spraying on fruit development time of 0.5% calcium chloride ($CaCl_2$) solution.
- High calcium limestone liming before planting 2-4 months can reduce blossom end rot.
- Increased levels of phosphate decreased the incidence of the disease.
- This obstacle is reduced by increasing the frequency of irrigation.
- By proper fertilization, water management and plant varieties tolerant managed to blossom blight.

2. Fruit cracking

Fruit cracking

Various categories of cracking injury have been described: radial cracking (Star shaped and initiating from the peduncle), fruit bursting, concentric cracking (Globular cracks initiating from the peduncle) and cuticle cracking also called as russeting. Cracking arises when there are amplemoisture and high temperatures, particularly when the existence of the subsequent situations prevails in these periods of stress due to quickgrowing. Fruit cracking is often associated with the rapid movement of water and sugar in the direction of the fruit when the elasticity and resistance of the cuticles are weak. In general, this condition occurred six to seven weeks after the set of fruit. In the fruit cracking phenomenon, an imbalance between water supply and water depletion has been implicated.

Management

- Irrigation at regular intervals.
- Picking of fruits before fully ripe stage which reduces radial cracking.
- Borax soil application @ 15-20 kg per hectare.
- In fruit 2-3 times the spraying borax 0.25% at maturity.
- Grow resistant varieties such as Sioux, Roma, Punjab Chuhara, Pusa Ruby, Arka Saurabh, Pant T-1.

3. Catface

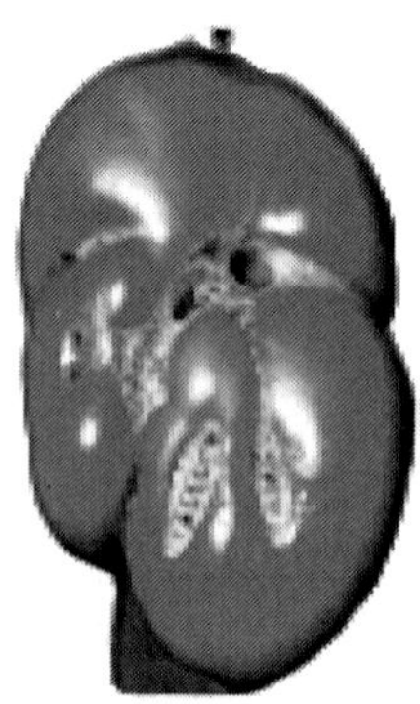

Catface

Catface is a term used at the end of the blossom and leathery scar tissue bands to define irregularly bulged fruit. Cold weather distorts and destroys those cells at the time of the blossom collection, which should turn into fruit, leading to deformities. Among the first fruit produced the condition is most commonly observed. Catface is most common in big fruit "beefsteak" type of tomato. Cat face abides in the first harvest.

Catface by the abnormal development of flower buds before flowering caused. This is often due to below 15°C prolonged cool weather when the plants are young. High soil nitrogen levels and excessive pruning can also create a shape error buds.

Management:

- Maintenance transplant production site of a temperature higher than 16°C
- Avoid high levels of nitrogen in the soil and excessive pruning.
- Avoid growing large fruit tomato varieties, if obstacles remain.
- Lack of water, which avoid the cycle.
- Adequate soil moisture balance maintenance.
- Recommended cultural practices should be adopted.

4. Sunscald

Sunscald grows on the sun-visible green tomato fruit. Whitish, lustrous blisters in the region are the preliminary indications. Killed, bleached tissue gradually collapses, and wrinkled and yellowish fruit ripening can develop into a slightly recessed area. Secondary organisms quickly attack the killed tissue and the fruit decays. Most of the fruit affected by sunburn are those who are suddenly exposed, because pruning, diseases caused by natural spread from a heavy load of fruit plant foliage or loss of sunlight. The degree of injury is more serious in the period of unusually high temperatures.

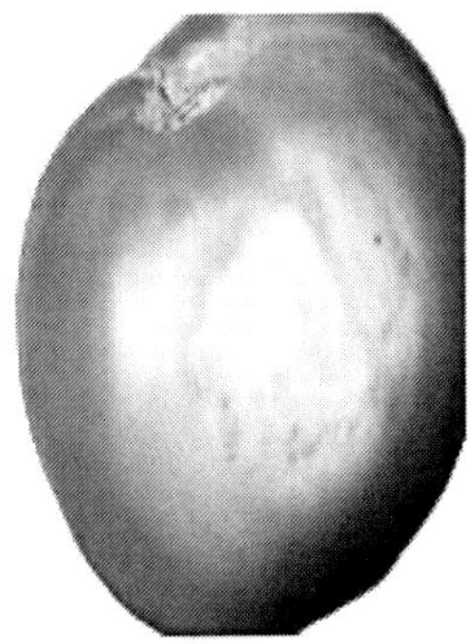

Sunscald

Management

- Avoid heavy pruning or shoot removal.
- Crops are planted at higher densities are susceptible.
- Prevent sunburn is to use the best protection with sufficient covering of leaf varieties of fruit, and provide adequate protection of water and pests, in order to maintain a healthy leaf.

5. Blotchy ripening

This physiological condition is a case where the normal red pigments of fruits local area indication. These areas show ripe fruit yellow or gray-green patches for other normal color. When these cut fruit, brown discoloration is often apparent. Climate, vegetation and cultural issues may cause spots maturity. Low levels have been associated with a disorder in the plants and or extended cloudy weather duration, potassium deficiency, insufficient light intensity. Other possible contributing factor is the high soil moisture, high humidity, low temperature, soil compaction and excessive fertilization.

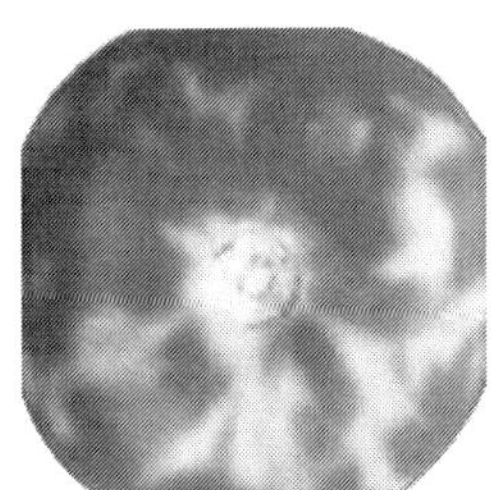

Blotchy ripening

Management

- Crop balanced fertilizer doses can prevent the occurrence of spots mature.
- Adjust planting dates to achieve good fruit development, good light intensity.
- In order to reduce the incidence of speckle mature, follow the correct cultural practices, in order to maintain nutritional balance and plant vigor.

6. Physiological leaf roll

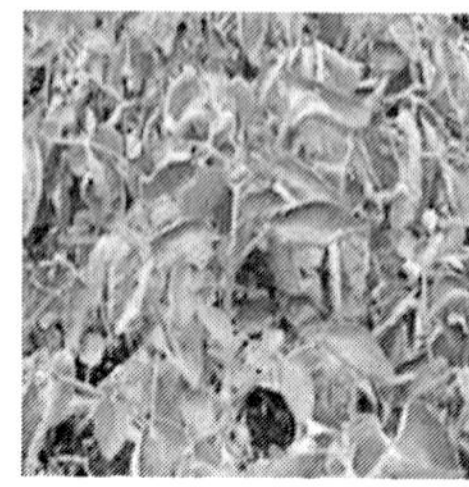

Physiological leaf roll

Rolling up the edges of the leaves and leaf physiological occur when inwardly. Sometimes curl continues until the contact or overlap from either side of the margin. Some of the leaves of plants may not exhibit the scroll. Leaves do not reduce the growth, yield and quality of plants and fruits. It is believed to be the result of irregular water supply, and can be improved after pruning. These symptoms are sometimes temporary and disappear after a few days, but they can last for the entire growing season.

Management

- Planting varieties determined.
- Well-drained soil and planting maintain uniform and adequate soil moisture.
- Be careful not to over-fertilize with nitrogen fertilizers and in particular to provide appropriate fertilizer.
- Avoid severe pruning.
- Maintaining the temperature below 95 °F or by evaporative cooling using a shadow.

7. Unfruitfulness

Temperature has a tremendous influence on the setting of the fruit on the tomato, especially at night. The fruit set is not suitable for higher days (> 32°C) and higher night temperatures (> 20°C). Fruits, however, are usually not fixed at or below 13°C. High and low temperatures impact fruit largely due to a decrease in pollen viability and a reduction in stigma pollen germination. That is why low fruit levels are a concern in eastern Indian summer crops, particularly in Western Bengal, where temperatures remain over the essential limit and when the temperature drops much below the crucial level in winter in northern India.

Management

- Hot set varieties such as: HS-102, Selection-18, Pusa-120, Pusa Ruby, Pusa Sadabahar, Pusa Hybrid-1, Punjab Kesri, Punjab Chuhara, can be grown.

- Cold set varieties such as: Pusa Sheetal and Pusa Sadabahar can be grown as low temperature tolerant varieties.
- Application of growth substances like Parachlorophenoxy acetic acid (PCPA) @ 50 ppm at full bloom stage or 2,4, dichlorophenoxy acetic acid (2,4–D) 1 to 2 ppm prior to anthesis of the first flower cluster and again 15 days after anthesis helps to set fruits during summer and winter seasons.

8. Herbicide injury

Herbicide injury

The disease is caused by abuse or 2,4-D, MCPP and other growth regulatory herbicides. Tomato plants are highly sensitive to these chemicals throughout the growing season. The first symptom is the growing stage and the tip of the blade is curled downwards the leaves are narrow and the torsade's show light towards the vein. The most obvious signs are that the plant part is actively growing these impressions. In severe cases, the trunks and stalks are thick, stiff, and easy with outer warts. However, the fruit may encounter cat or plum-shaped growth, may be hollow and seedless.

Management

- Do not spray when the wind will carry the spray drift towards sensitive tomatoes or other crops to prevent herbicide damage.
- Furthermore, spray at low pressure with a coarse spray nozzle and spray as close to the ground as possible.
- Avoid herbicides have previously been included because traces of herbicides, even after thorough rinsing may remain in the sprayer applications of pesticides.

9. Goldspot or Goldspeck

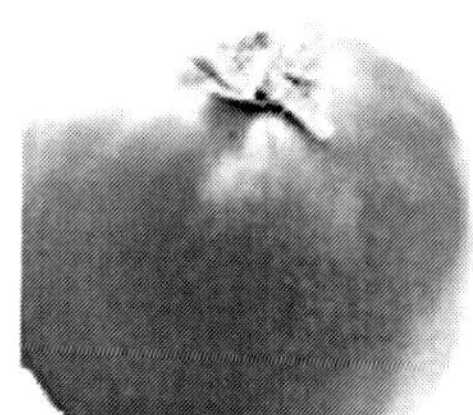
Goldspeck

Gold spot or spots often observed on the calyx and fruit ripening shoulders around, especially in the summer. In green fruit, white spots, not too rich. These spots less attractive fruits and significantly shorten its shelf life containing fine crystals of calcium salts, probably calcium oxalate granular mass cells.

Management

- Varieties tend to be more sensitive to the gold spot on the BER resistance.
- Increase nutrient solution conductivity reducing the incidence of spot gold.

10. Russeting (Cuticle Cracking)

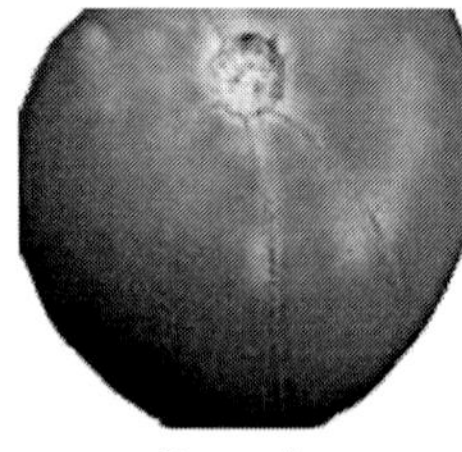

Russeting

Russeting is a tomato skin disorder in which 25% of the fruit surface is covered by hairline cracks that are not visible to the naked eye for minutes. The rain check, crack, swell cracking, shrinkage cracking, and cracking the spots of hair cuticles are also called. When the apparently crazy check carefully, and not smooth, the fruit has a rough feeling. It is sometimes described as poor skinned fruits and significantly reduces the shelf life of the harvested fruits. Six to seven weeks of fruit set up cracks in fruit ripening practices and the second highest incidence of the disease is early production and late season.

Management

- Select disease-resistant varieties.
- Avoid large fluctuations between day and night temperature and relative humidity.
- Avoiding major changes in nutrient solution and EC 3.0 conductivity is kept to a minimum in the minimum rock wool slabs and deleafing process.
- During the harvest by protecting leaves from disease and injury to avoid direct exposure to the fruit in the sun.
- The solution, increasing the calcium ratio by adjusting the leaves and fruits.

11. Edema (Oedema)

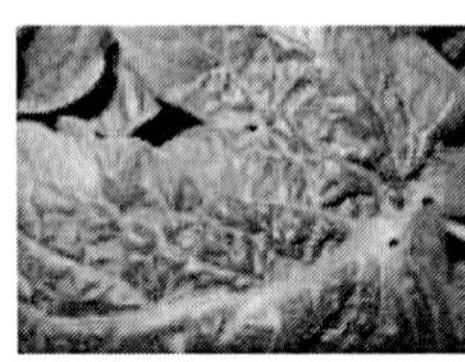

Edema

Callus growth swelling type image leaf blister similar to undifferentiated in its early stages. Fresh grainy appearance blisters caused by pressure splitting of the epidermis from within the potential. This exposes the parenchyma cells to swelling. Eventually, these swollen parenchyma cells erupt. In a period of time results in distortion of these cells rupture and distortion of leaves and necrotic regions, when the dry cell.

Management

- The condition can be prevented or remedied in a greenhouse or growth chamber.
- Reduce watering, increased ventilation by the following measures, higher temperatures and higher light promote transpiration should be effective.
- In the field, the disease occurs only when there is an excess of water and low transpiration in the long term. Another point can be achieved in the field, assuming, in addition to trying different varieties that irrigation has stopped.

12. Puffiness

Puffiness

If the outer wall is normal, but inside the tomato is hollow, where the seed cavity is usually empty. Reason: high or low temperature limits, excessive nitrogen fertilizer, and heavy rain may normally pollination interference, resulting in swelling of the fruit. Puffiness occurs most commonly on early fruits.

Management

- Over irrigation should be avoided.
- In boron deficient soil condition, boron should be applied by spraying 0.3 to 0.4 % borax solutions at initial plant growth, fruiting and fruit development stages.
- Less nitrogen should be applied to crops in short days and low light intensity conditions of growth, which swells rather than long days, a high light intensity.

13. Vascular Browning

Vascular browning and brown phase tissue killed at a later stage. Browning of vascular system near the cloud appears to be characteristic of polyphenol oxidase browning. Polyphenol oxidase has been found to be abnormal wall tissue mature cloud effects active, but is inactive in normal or even the impact of cloud computing, fruit ripening wall tissue. High polyphenol oxidase activity has been found in very young fruit, reduce and eventually disappear with maturity increases. It was suggested that cloud affect fruit display polyphenol oxidase activity blotches regional anomaly reserved. Clouds tissue still exists in this activity, ripe orange-red stage on plants.

Management

- Foliar spray of 1.5 % Magnesium Sulphate, when symptom appears.

14. Zebra Stripe

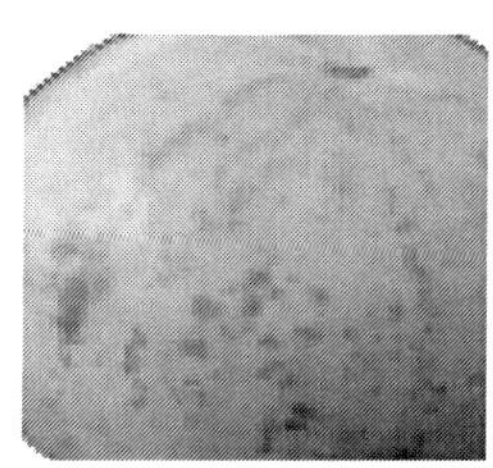

Zebra Stripe

Zebra can be characterized as a series arrangement from the stem end to the blossom end of the line dark green spots. It can lead to elongated spots coalescing tag. This problem is due to genetic control.

15. Zippering

Zippering may be described as having a thin scar, or fruit from a fully extended scar area to flowering stem end portion. Occasionally locules opening accompany zipper scar. Some people believe that when fruit and flowers glued to the abnormal shedding, but this may be related to the proper fruit development and the formation of damaging interference zipper.

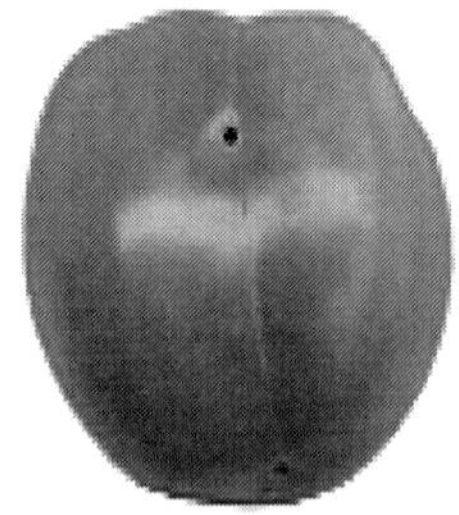

Zippering

Management

- Good cultural practices, avoiding fruit emphasized.
- Use varieties can be difficult to alleviate the problem of zippering.

B. BRINJAL

1. Poor fruit set

Four flower groups related to style length are born in eggplant, independent of the variety *viz.*, long style, medium style, pseudo short style, and short style. In general, pseudo short and short-style flowers do not produce fruit, but their number in a plant is typically greater than fruit-producing long and medium-style flowers. Flowers in brinjal are commonly assisted alone as well as in clusters. Solitary flowers are typically long or medium in style, while the shortcuts are in the cluster mix style, short and medium/long style flowers.

The ratio between short-style flowers and long/medium-style flowers in the group is apparently very high (3-4:1). So, the collection of fruits depends on the number of viable flowers in the plant (long/medium style).

Eggplant is also the natural decline of a problem may spend up to 50% in the medium style and 30% in long-style flower especially under high temperature conditions. Unproductive flowers and drop of flower advantage sometimes result in weak eggplant.

Management

- With 2ppm of 2,4, D spray flowering plants in the small flower clusters appear.
- In the full bloom stage of plant spraying 60ppm of NAA or 50ppm of PCPA.

2. Blossom end rot

Blossom end rot

It showed the general decay in the development of fruit blossom end. In order to slow down the movement of water, insufficient or uneven irrigation of plants, high humidity or the result of other factors is usually the plant. Because calcium is taken up by plants in the transpiration stream (water), water often leads to a slow-moving flower end rot with a temporary lack of calcium.

Management

- Increase in the frequency of irrigation.
- Regulation of soil pH by liming
- 0.2% of calcium chloride foliar sprays at fruit development time.
- 40g CaO soaked in one liter of water overnight and then mixed with 10 liter of water and sprayed twice or thrice.

3. Calyx withering

This physiological disease generally occurs between mid-February to April. Harvest before symptoms of the disease become apparent seven days. Affected by the color of the fruit, its lack of normal luster, thus reducing the marketability of the fruit becomes reddish. Calcium and nitrate contents are much higher in affected fruits than healthy ones.

Management

- Spray @ 0.1% borax and place a rubber band around has begun opening its calyx of the flower.
- low level of boron enhances calyx splitting
- Closer spacing
- Irregular or fluctuating temperature

C. CHILLI AND CAPSICUM

1. Flower and fruit drop

At that time the high temperature, rainy summer, early summer, blowing hot air and moisture susceptibility factors in addition to the virus is to limit pepper cultivation of high temperatures and lack of humidity is too large. Such conditions increase the flowers and fruit drop and reduced fruit size, fruit number is reduced the yield.

Management

- Irrigation during flowering and fruit setting stage to reduce the flowering and fruit drop.
- Effectively control the spraying of the foliage of flowering NAA 50ppm decline.

2. Sunscald

Sunscald

Sunscalds often occur when the plants are prone to leaf disease or quick pruning of the plant's canopy. Exposure to high intensity can lead to tissue injury after attaining mature green phase. Initially the exposed side of the sun becomes white because the chlorophyll then disappears and is converted to black. Exposure to only 10 minutes of strong sun exposure to cause such damage if the tissue temperature rises to sun exposure of 50°C. The affected areas are light-colored, soft and wrinkled. The damaged tissue eventually turns to a pale brown papery texture Sunburn is often followed by secondary fungi and bacteria, and can help with further decay. Bell Pepper The disease was found to be more sensitive than pepper. High temperatures lead to the presence of sunburn. Fruits become mature when the disease is more sensitive Color changes from green to red. Disease control leaves with pesticides, to prevent leaf drops, which expose the fruit to sunlight or to sun-shaded frame marks.

Management

- Make available proper water and fertilization to produce enough foliage stronger plants.
- Grow about 12 inches apart in two rows of pepper. The line will cultivate together and protect fruits.

3. Blossom end rot

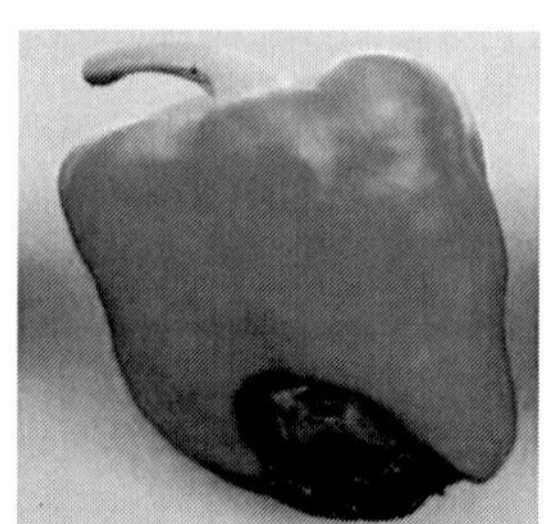

Blossom end rot

Developing fruit show rotten at the blossom end. Development blossom end rot fruit display. It is basically due to the lack of irrigation, or other factors unevenness, high humidity, or by slowing the movement of the vegetation water. Because the calcium is taken into the transpiration stream (water) plants, water often leads to a slow-moving blossom end rot temporary lack of calcium.

Management

- Careful and appropriate irrigation.
- Phosphorus, potash sensible amount applied with nitrogen fertilizer.
- Two kinds of foliar application of 0.2% calcium chloride at fruit development time.

D. POTATO

1. Blackheart

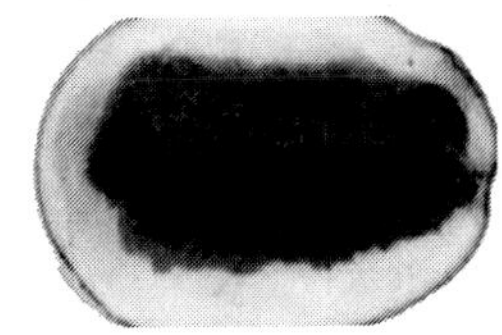

Blackheart

Which often sinister dark gray to black is in color indicating the central necrotic tissue tubers first visible. Affected tissue is firm to the touch, if you can smell the mellow warmth. Dimmable cavity region, which is usually formed from the interior surface of the separation. However, black heart symptoms can reach the surface appears as patches of deep depression and infections are likely to be destroyed on the surface.

Management

- Avoid excessive drain on good soil and excessive irrigation.
- Prepare the soil structure and porosity to make air circulation.
- Avoid harvested in extremely hot (> 30°C) or low temperature (around 0°C).
- Maintain moderate temperatures during cold storage.
- Tuber dry quickly, in order to avoid condensation may reduce air exchange.
- Maintenance during transport and storage in adequate ventilation to avoid suffocation tubers.
- Avoid storing tuber in tightly closed places.

2. Tuber cracks

Tuber Cracks

May result in irregular growth conditions during the tuber can burst open surface cracks and skin tissue tuber expansion. Depending on the source, type and intensity of the fracture, symptoms may vary widely, from the roughness of the skin-deep tubers or more fractures in the same individual. The main reason is the development of deep cracks in irregular tuber expansion of irrigated crops. Following irregular growth, due to drought, watering will lead to rapid infusion vascular tissue. The resulting internal pressure will be higher than the resistance of the skin, healing and greater reaches this.

Management

- Do not planting susceptible varieties.
- To ensure good rooting good soil preparation.
- Regular and proper irrigation is conducive to the development of even tubers.
- Reasonable amount of nitrogen.

3. Pink pigmentation of the flesh

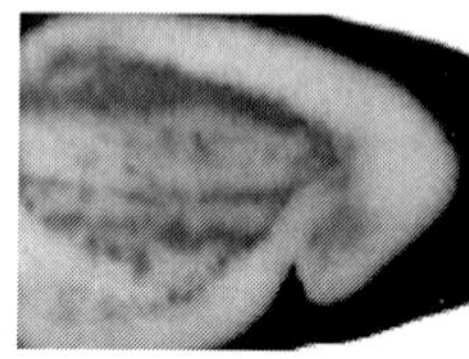

Pink pigmentation

Two types of pink colouration may be observed in the flesh of the tubers:

- Purplish pink colouration is very clearly defined and localized at the vascular ring and takes the form of an unevenly shaped area
- The pink colour is more or less intense and takes the form of mottling, usually localized in the pith of the tubers but sometimes taking up virtually all of the volume of the flesh.

Additionally, pink colouration of the skin may also be observed externally in several yellow skinned cultivars which are to varying degrees also susceptible to this anthocyanic pigmentation. This may also affect the eyes or possibly the entire surface of the tuber. Pigmentation of the eyes may, however, be a genetic characteristic of some cultivars.

Management

- The extent of the symptoms of pink flesh pigmentation can be limited by preparing large ridges and by frequent, but not excessive, irrigation.
- As regards skin pigmentation, tubers should not be exposed to daylight except during the handling operations required for harvesting, storage and packing, and to as little as possible direct sunlight.

4. Greening

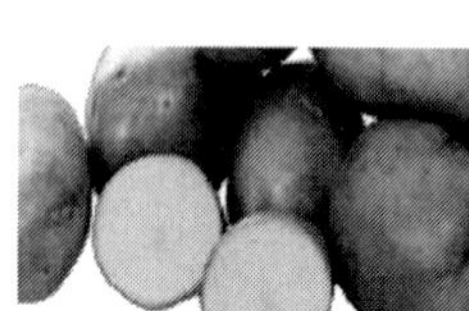

Greening

Green potatoes are result of too much exposed light, which triggers the tubers to overproduce solanine alkaloid (a toxin substance). More than 14mg/100gm of solanin makes potatoes bitter in taste and very toxic consequences if consumed by human and cattle. From the originating stolons, tubers having all of the features, including a light, thus becoming green capacity under action, to produce chlorophyll. If the tubers in the soil development, but if economic growth during storage or exposure to light may occur afforestation does not occur. Green superficial tubers usually more intense than the diffuse green after harvest. The pigment synthesis is substantially located in the first 3 millimeters of surrounding tissue tubers. Solanine is the defensive mechanism of plant adopted for potato tubers to fight pests and diseases like early blight.

Management

- Prepare the soil in such a way as to avoid the formation of clods and adapt according to the type of soil tillage depth.
- Ensure the spacing between rows 90 cm, and the depth of planting varieties adjusted.
- Prepare the soil with excellent quality wide ridge, keeping seed ridge center.
- Irrigation to prevent the formation of cracks and ridges damage.
- If necessary, fill cracks in the soil by "rolling", after the destruction of haulm.

5. White flesh and immature tubers

Internal

When the tuber is cut open the colour of the flesh is seen to be very pale, usually white but not glassy. This pale colouration may concern a whole potato lot, a batch or sometimes only a few tubers, or even only parts of the tuber.

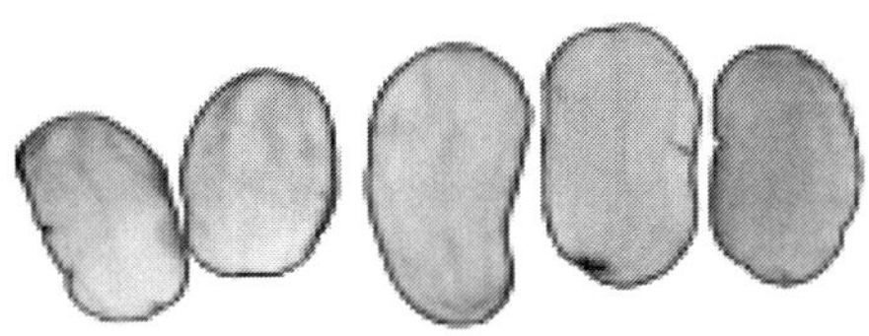

White Flesh

External

White fleshed tubers sometimes have a finer skin, sometimes even loose skin and are more prone to rot, wounds and thumbnail cracks on the surface of the tuber.

When seed tubers are affected, their ability to germinate may be diminished, particularly in cold conditions in spring and irregular emergence of the plants may occur. If hot conditions occur during the growing period, the seeds may rapidly liquefy and weaken the plant which will then be more susceptible to disease.

Management

- Prevention basically consists of reaching adequate tuber maturity before haulm destruction and cultivation methods that reduce growing stress (appropriate fertilization and irrigation, etc.).

6. Brown heart and hollow heart

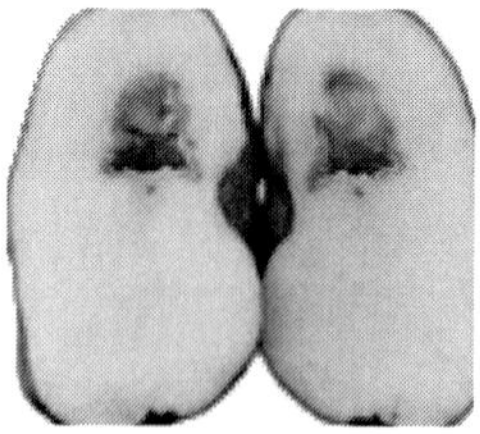
Brown heart

Brown heart starts with a brown necrotic spot inside the pith of the tuber. When very extensive, it affects virtually the entire tissue and a crack form in its centre. However, the tissues do not always crack and it may simply remain as a brown spot.

Hollow heart manifests itself by a longitudinal cavity usually around the centre of the pith and sometimes situated near the rose end or the heel end. It is sometimes cross shaped but its shape can also be irregular and take up almost the entire volume of the pith. The walls of the cavity become brown due to the formation of healing cork. Hollow heart usually appears in the centre of large tubers. In conditions of slow and regular growth, necrosis normally develops very little and remains at the brown heart stage, whereas in adverse conditions the tubers enlarge too rapidly causing the medullary (pith) tissue to rupture.

Management

- Do not plant too early, especially if the spring is cold and wet.
- Use moderate fertilization.
- Ensure a good tuber initiation by keeping the soil moist during this phase.

7. Enlarged lenticels

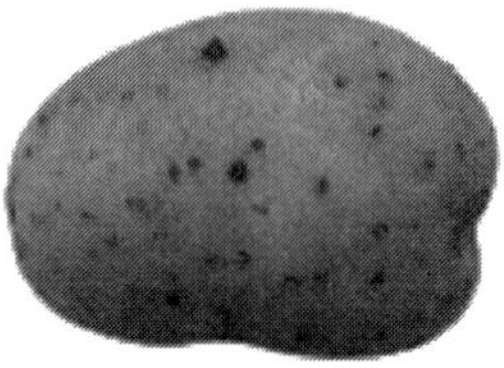
Enlarged lenticels

When the tuber takes a long time to form under wet conditions in the soil to develop enlarged lenticels. Wherein the first drying them, and then to form a brown suberize point measurement is 1 to 3mm many white grow. These are generally circulated on the surface of the tubers.

Management

In the field and harvest management

- Hold 90 percent below the usable soil moisture, if possible.
- The harvest temperature is among 50-65 °F.
- After heavy rains or floods, prevent harvesting.
- Cleaning and disinfection harvesting tools.
- Reduce bruising.
- Suberize tubers correctly.
- Never wash tubers prior to storage.

Storage management practices

- CO_2 accumulation and condensation prevention storage, to ensure good air flow.
- Regular cleaning and disinfection fluming system and water.
- Efficient fungicides to clean water.
- Clean and disinfect all equipment to prevent the spread of germs.
- Dry quickly tubers with forced hot air to remove water film tubers.
- Ensure good ventilation bags and containers.

8. Secondary growth: deformed and glassy tubers

The shape of the tubers depends on when climatic conditions change during the growing process.

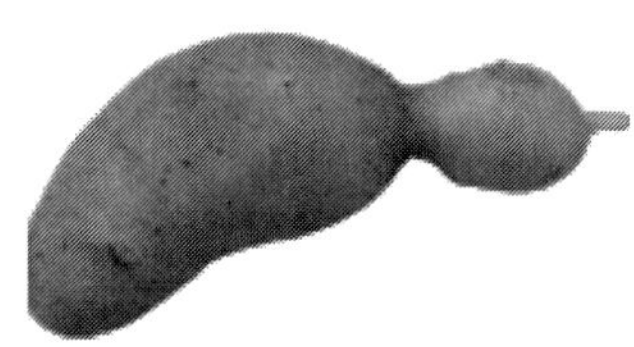

Secondary growth

- When growth cessation occurs early in tuber development, it induces contraction at the base end (pear-shaped tuber)
- When it occurs, the two parts formed during tuber development may be of similar size (dumbell-shaped tuber or ridge roller)
- Fecal growth at the tip of the rose, which occurs at the end of tuber development (this growth also occurs in many lateral eyes, providing protrusions of various shapes and sizes (tubules with multiple handles).

Management

- Vulnerable breeds should be avoided in difficult conditions. In general, early varieties and varieties with round or short rectangular tubers are less susceptible to elongated tubers.
- Good cultivation practices: avoid irregular growth, good soil structure, adequate density seeds, attention to ridge volume, sufficient nitrogen input and irrigation. Also, studies have shown that the application of maleic hydrazide for many years favorable to this phenomenon may partially limit secondary growth and formation of vitreous tubers.
- 2nd generation of tubers arrive, kill the stem. This will be small enough when it comes to grading that can be removed. If carrier destruction is delayed for harvesting reasons, the time the tubers spend in the soil should be kept to a minimum to reduce vitreous.

9. Internal rust spot, internal brown spot

The phenomenon of rust spots (also called physiological rust) is characterized by the presence in the flesh of the tubers of round necrotic areas. These are rust-brown and of irregular shape and size. Whilst these lesions may be localized in all the internal area of the vascular ring, they usually appear more intensively next to the rose end, in the perimedullar parenchyma. They do not develop further during storage. Although, these symptoms are quite common in potatoes grown in sandy soils.

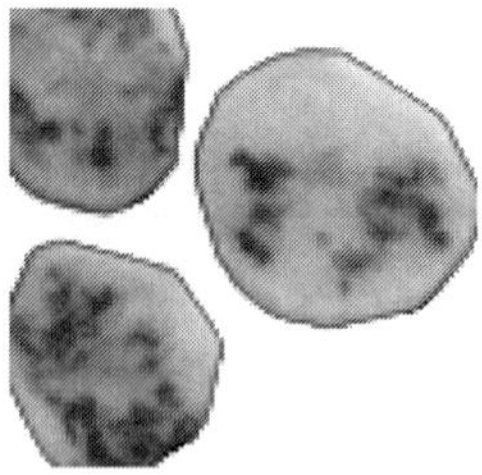

Internal Brown Spot

Management

- It is advisable to avoid susceptible cultivars in regions of high-risk soil.
- The extent of these symptoms can be limited by large ridges and frequent, but not excessive, irrigation to prevent the tubers from enlarging too quickly.
- On planting, the application of calcium in a readily absorbed form may also prove effective and a number of studies have pointed to a reduction of the symptoms by the use of maleic hydrazide.

4

Physiological Disorders of Cole Crops

A. CAULI FLOWER

1. Browning

Browning

Browning is due to boron deficiency caused. In general, boron deficiency symptoms in the formation of the curd, the plant externally visible. In the early stages, immersion in water and curd in the stem region of the surface occurs. As plant growth, into the hollow, soaked in water covers the cavity inner wall tissue. In the absence of progressive steps, areas of brown or pink color are visible, and hence, it is also called brown or reddish-brown or dark brown rot affected by the bitter taste. Dark to dark green and light-yellow leaves on the upper surface of the old leaves. When boron is severely deficient, then leave less to develop and shrink.

Management

- By adding borax, boron deficiency may be rectified. Borax depends on the type of soil, the PH value of the soil and the degree of insufficiency.
- Borax 10-15 kg/ha is adequate in acidic soils, while larger quantities, such as natural and alkaline soils, may be needed.

2. Whiptail

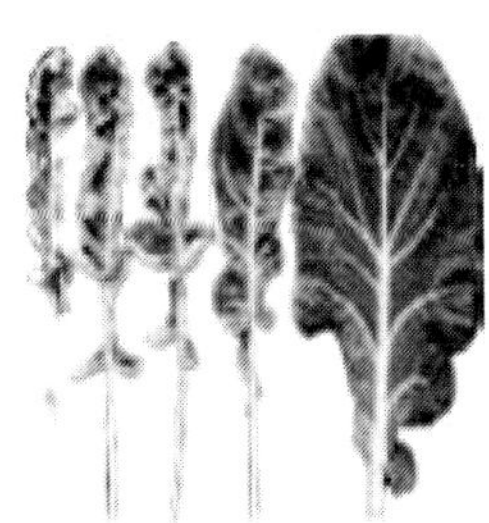

Whiptail

Whip-tail disease is caused by a lack of molybdenum, and the deficiency symptom of young plants is leaf marginal bleaching, which can cause the entire leaf to turn white, and if the leaf is not normally developed, it is often mentioned when there is a serious deficiency only in vein development with “Whip Tail”. The plant’s growth point is also modified to prevent the development of tofu, and when the soil pH is less than 5, less molybdenum is generally produced in acidic soils.

Management

- Lime to raise soil pH to 6.5. The use of lime in acid soils is to increase molybdenum supply.
- Alternatively, soil application of sodium molybdate (10-15 kg / ha) the lack of effective control of symptoms.
- Spray with 0.1% ammonium along with three stickers also gives satisfactory results.

3. Buttoning

The production of tiny premature curds or buttons is called buttoning, and these plants are young. Exposed button head, display conditions, the plant leaf has a generally small underdeveloped. There are several factors, such as poor supply of nitrogen, over the age of planting seedlings, planting and adverse weather conditions caused by improper time report button.

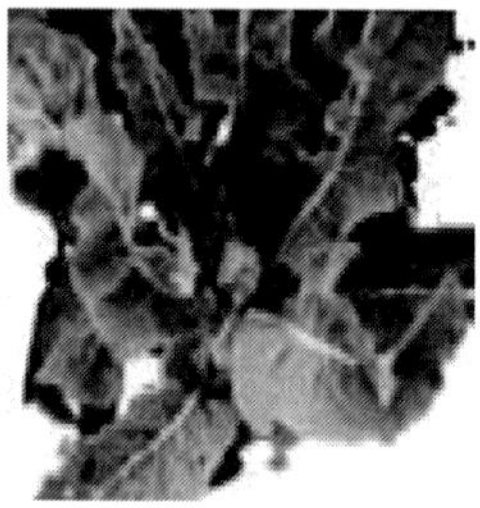

Buttoning

Management

- Nitrogen and moisture for growing plants in nutrient supply fast enough to be considered very important to prevent the occurrence of plant button.
- To be transplanted seedling age should not be more than five weeks.For the prevention of the occurrence of button plants, an adequate supply of nitrogen and moisture for rapid vegetative growth of plants is considered necessary.
- Right selection of variety.
- Age of the seedlings to be transplanted should not be more than 5 weeks.

4. Blindness

Blind cauliflower plants are those that lack the terminal bud. In cabbage and Brussels sprouts, the disease exists. This results in a plurality of small heads in the head, or in the form of no. There are many causes, including pre- and/or post-transplanting, such as the damage of the terminal growing point due to low temperatures, damage to the cutworm, or rough handling of transplants.

Blindness

Management

- Plants should be avoided exposure to very low temperatures.
- Careful handling of plants and good plant protection measures, so that the terminal bud may not be injured

5. Bursting/splitting of heads

Bursting

Heavy rain typically follows split cabbage heads, particularly after a time of dry weather. The head from the split is triggered by increasing the internal pressure after the excess moisture absorbed by the roots cabbage head firmly. Fertilization could happen the same thing when the head finishes the season. Maturing varieties are more susceptible than varieties of late split cabbage,but all varieties can be broken down under suitable conditions.

Management

There is no simple fix to split the cabbage so prevention is very important. Here are some things you can do to prevent the cabbage head split:

- Throughout the growing season keep the soil evenly moist.
- Cabbage needs a week 1.5 inches of water, either as rainfall or supplemental irrigation.
- Cut some of the roots when the head is close by culturing plant with a hoe modest company. Trim the roots of plants can reduce the amount of moisture absorption and prevent split cabbage.
- Avoid strong head start after fertilization. Use slow-release fertilizer can help keep the nutrient levels of the soil, and even prevent excessive fertilization.
- As long as the firm head is harvested early variety.
- Plant cabbage early to ripen before the next set of warm environments. This can be done as early as four weeks before the last frost.
- Use transplants instead of seeds to crop a good start.
- In the field of a short spring, the growth of Chinese cabbage crop fall. Fall plants about 8 weeks before the first frost expected crops.
- Use of organic mulch to help the soil retain moisture and keep the roots cool.
- When the cabbage head to prevent the split despite its best efforts, split the first harvest as soon as possible. Splitting the first header stores not only to solid, so the first header using a split.

6. Tip burn

The characteristics of the disease in the leaf edge, edge tips or cabbage, Brussels sprouts and cauliflower necrosis. The development of brown spots in the tendency to decompose during transportation and storage, and to enter into speculation decay-causing organisms. Problem is caused by calcium deficiency; grow rapidly due to excessive nitrogen, high temperature and lack of water. Tip burn is generally regarded as CA-related disorders, causing localized calcium deficiency leaves or leaf edges, but it actually seems to be quite a stress related disorder. The problem is that even in the presence of calcium translocation into the soil in sufficient quantities to accommodate a rapidly growing one plant. Nitrogen and nitrogen supplementary applications especially applied in the form of ammonium nitrate with respect to the form should be timed to avoid the rapid growth in the later stages of plant development.

Tip burn

Management

- The disease can also ensure adequate root growth by the operation less sensitive to cultural practices and the use of varieties to avoid.
- Growing resistant varieties.
- Stop unnecessary fertilization and use a conditioner to increase the available calcium in the soil.
- Calcium salt foliar nutrient solution.
- Irrigation control helps regulate the growth of plants.
- Avoid unnecessary fertilization and increase the calcium available in the soil by changes in the soil.
- Apply calcium salt solutions with foliar nutrients.

7. Leafy curds or Leafiness

Such situations are found between the portion of the green leaves and curds are considered by the relatively high temperatures and delayed harvest caused. Irregularly shaped loose curd and the curd cannot be sold due to losses to growers.

Leafiness

Management

- Selection of right variety for the particular growing condition.

8. Riceyness

Premature initiation of floral buds on the upper surface of curd is called as riceyness. Rendering curds become granular, loose and poor quality for marketing.

The loose surface curd velvety appearance due to pedicels and are formed with small elongation in the curding flower bud stage, e.g. curd called ricey.

Management

- Correct selection of varieties and transplanting at the right time.
- Good seed stock and favourable weather condition tend to prevent this disorder.

Riceyness

9. Fuzziness

Flower peduncle elongation becomes soft and smooth and curd elongates. The variety of the disease is encouraged to cultivate their normal growing season. In appropriate cultural practices in quality seed sowing the right time to reduce fuzziness.

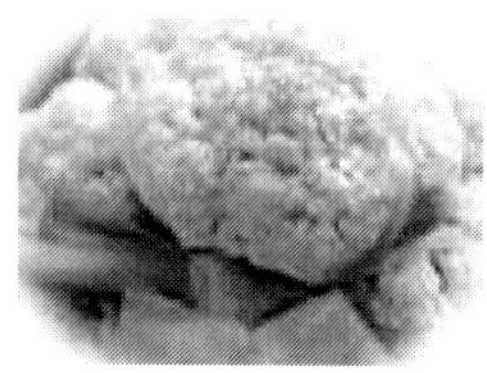

Fuzziness

10. Hollow stem

When curd hollow field, due to the remainder of the curd below, extending central core or meat division uneven growth rate. To create a cavity in an external setting, the cavity can extend to either end of the plant. The most serious cases are rising rapidly in this situation. Several explanations have been identified for the hollow stem, including environmental stress. Suddenly, the rapid growth is often irregular, N high grade and low plant population is in favor of the hollow stem conditions. It has also been documented that boron deficiency is associated with this disease.

Hollow stem

Management

- Optimum use of nitrogenous fertilizers.
- Application of borax 1.1 to 3.4 kg per hectare.
- Increase planting density, to avoid excessive soil fertility, and to encourage their conditions of reduced growth rate of plants.

B. CABBAGE

1. Blindness

In cabbage and kale, the disease happens. This arrangement results in no small heads in the head, or in the form of a plurality. There are many causes, including pre and/or post transplanting, such as the increasing point of damage to the terminal due to low temperatures, damage to the cutworm, or roughened movement.

Management

- Plants should be avoided exposure to very low temperatures.
- The terminal bud may not be injured so that careful handling and good plant protection measures should be followed.

2. Bolting

Hot weather is usually the cabbage's trigger for bolting, but sudden exposure to low temperatures at night may also cause plants to go to seed. A lack of moisture or nutrients is other potential stressors. Once plants have begun to bolt, there's no way to stop the process. If cabbage is detected in early stages of bolting, the flower and stalk may be nipped off to buy a few extra days of harvesting. Even if the flower is removed, the edible leaves, stems and roots soon turn tough and bitter.

Management

- Timely sowing check for this disease

3. Bursting/splitting of heads

Head usually split or burst due to heavy rain or excessive irrigation or cabbage delayed harvest. Rapid growth has led to the first division. Early-maturing varieties most likely to be affected. Such disorder may be excessive watering and maintain proper irrigation alleviated by avoiding uniform growth. In late varieties best time to harvest and use of resistant varieties may help alleviate the problem.

Management

There is no simple fix to split the cabbage, so prevention is very important. Here are some things you can do to prevent the cabbage head split:

- Throughout the growing season keep the soil evenly moist.

- Cabbage needs a week 1.5 inches of water, either as rainfall or supplemental irrigation.
- Cut some of the roots when the head is close by culturing plant with a hoe modest company. Trim the roots of plants can reduce the amount of moisture absorption and prevent split cabbage.
- Avoid strong head start after fertilization. Use slow-release fertilizer can help keep the nutrient levels of the soil, and even prevent excessive fertilization.
- As long as the firm head is harvested early variety.
- Plant cabbage early to ripen before the next set of warm environments. This can be done as early as four weeks before the last frost.
- Use transplants instead of seeds to crop a good start.
- In the field of a short spring, the growth of Chinese cabbage crop fall. Fall plants about 8 weeks before the first frost expected crops.
- Use of organic mulch to help the soil retain moisture and keep the roots cool.
- When the cabbage head to prevent the split despite its best efforts, split the first harvest as soon as possible. Splitting the first header stores not only to solid, so the first header using a split.

4. Tip-burn

The characteristics of the disease in the leaf edge, edge tips or cabbage, Brussels sprouts and cauliflower necrosis. The development of brown spots in the tendency to break down during transportation and storage, and to enter into speculation decay-causing organisms. Problem is caused by calcium deficiency, grow rapidly due to excessive nitrogen, high temperature and lack of water. Tipburn is generally considered a disorder related to Ca, causing localized calcium deficiency leaves or leaf edges, but it seems to be quite a stress related disorder. The problem is that even in the presence of calcium translocation into the soil in sufficient quantities to accommodate a rapidly growing one plant. N and N supplementary applications especially applied in the form of ammonium nitrate with respect to the form should be timed to avoid the rapid growth in the later stages of plant development. The disease can also ensure adequate root growth by avoiding the use of cultural practices and varieties is not easy to manipulate. After Tipburn may be secondary to opportunistic infections or rot pathogen Erwinia spp.

Management

- The disease can also ensure adequate root growth by the operation less sensitive to cultural practices and the use of varieties to avoid.
- Growing resistant varieties.
- Stop unnecessary fertilization and use a conditioner to increase the available calcium in the soil.
- Irrigation control helps regulate the growth of plants.

5. Black speck or pepper spot

This condition occurs in the outer leaves of the head with varying degrees of intensity, but can always be seen well into the middle of the head. In the field or after harvest, symptoms may occur. Black spots become apparent after a period of refrigeration. "Pepper spot. The exact cause of hot pepper spot is unclear. However, the G x E combination of factors may affect the form of this disease is good or bad.

6. Black petiole or black midrib

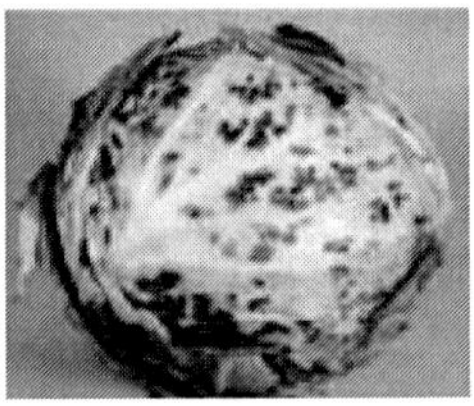

Since the head is approaching maturity, black at the dorsal side at a point or near the internal leaf petiole or attached to the core. Affected areas may extend under magnification found in collapsing the thin-walled veins cytoplasm along 5 to 7.5 cm. However, cells of vascular discoloration are usually free. The disease is believed to be a low level of Potash and high level of Phosphorus. K-P concentration imbalances and associated results in the soil. A relatively high nitrogen level aggravates the problem. Gx E interaction plays a significant role. The disease is manifested after a period of storage.

Management

- The correct combination of P and K and the avoidance of excessive nitrogen will prevent the condition of the black petiole.
- The suggested amount of K is 200-300 kg/ ha and P is 25-40 kg / ha.
- Manipulate resistant varieties and the use of the environment may provide solutions.

7. Vein streaking necrosis

The scale of the necrotic lesion may be close to the pepper spot on the surface of the blade near the midrib of the leaf. The problem's origin is unknown. Symptoms similar to pepper spots and streaking vein necrosis have been shown to create certain viral illnesses.

8. Necrotic spot

Feature of this disease is similar to pepper spot lesions. Cell death and pepper spot group at the same time, they are often associated with vascular tissue, rather than pores. Unlike intravenous streaking, necrotic spots affect the parenchyma cells, and epithelial cells unaffected. Symptoms of the blade or oval spots in the veins of the recess. The disorder usually developed during storage, and no symptoms shown in harvest. Preferably small impact cultivars with disease. Compared to susceptible varieties, refrigerated storage when the disease becomes controlled atmosphere (CA) storage is poor.

C. BROCCOLI

1. Buttoning

This condition characterises the formation of little unmarketable curds called buttons. Small plant buttons are developed with small leaves without covering the upward head. Among other things, the cause of this disease is inadequate nitrogen (N) supply to prevent transplanting. Inadequate water or other pressures for plant development as causative variables were indicated. Any growth restriction by dry soil or delay in planting might lead to the formation of little curds. There have also been reports of excessive hardness or too much hardening, cold temperatures at 4 - 10 °C for 10 day or more, illness, insects and vitamin deficiencies, which induce broccoli and coagulant buttoning.

Management

- This condition can be alleviated by the availability of sufficient quantities of N and timely field transplantation.
- Minimizing transplant shock and optimizing factors that favour vegetative growth can decrease the disorder's incidence.
- It will mitigate the buttoning issue by using younger, smaller transplants and cultivars with longer juvenile periods.

2. Hollow stem

In certain parts of the world, this disease exists. It occurs in cauliflower and broccoli. When the heart or fleshy core splits with the rest of the plant due to an uneven growth rate hollow area stretch from below curd. In order to create a cavity for the outside environment, the cavity can extend to either end of the plant. Infection by bacteria and fungi is possible and spoilage follows. When growth is rapid, this condition is most extreme. A variety of hollow stem causes, including environmental stress, have been identified. Sudden and rapid, typically erratic, high N levels and low populations of plants are conditions that favour hollow stem growth. Boron deficiency, while it is not critical, has also been reported to be associated with this disease. The occurrence of hollow stems in subsequent broccoli spears was increased by the additional application of K to broccoli transplants. Curds are unmarketable with hollow stems. From excessive N fertilization, hollow stems may also arise. The cause of this condition is deficiency or unavailability of boron in limed soils. Boron-deficient plant leaves are often blistered and yellow in colour.

Management

- Optimum use of nitrogenous fertilizers.
- Application of borax 1.1 to 3.4 kg per hectare.
- Increasing the density of plants, avoiding excessive soil fertility, and fostering conditions that reduce the rate of plant growth.

5

Physiological Disorders of Bulb Crops

A. ONION

1. Poor seed germination

Poor seed germination

In general, germination of onion seeds is poor, and, in the case of old seed stocks, this problem is severe. Seeds soon lost viability in the ordinary storage state. However, cold storage conditions significantly increase the viability of seeds.

Management

- Store in cold storage the seeds.
- Using good seeds and new ones.
- Soak the seeds for 2-6 hours in water and sodium phosphate solution, drying is followed, which is good for the vigour and viability of the seeds.

2. Bolting

Bolting

It's a serious onion illness. It refers to seed stalk production before its inception and has a detrimental influence on bulb generation and development. Bolting is unwanted, as it impacts the output of the bulb directly but is good for seed production. The bulb is brilliant and fibrous, and storage of these bulbs does not survive long. Rabi onion requires a moderate temperature of 20-25°C (December - May) and a lengthy photographic time will take place in high-temperature circumstances for bulbing and bulb development. On the other hand, Kharif onion is raised in a somewhat shorter day and a mild temperature condition (August to December).

Management

- Adjust transplant timing to allow seed to expose moderate bulb temperatures. In relation to the kharif crop, the maturity of the rabi is correlated with the high temperature.

- The *Kharif* onion planting at the beginning of August and the *Rabi* onion plantations at the end of December to January produce less bolters.
- Growing of cultivars without bolting like Early Grano, Texas Early Grano, etc.
- Solid and eight weeks old seedlings transplanted.
- Provide the dosage of nitrogen recommended.
- At an early stage, the seed stalk is removed.

3. Splitting and doubling of bulbs

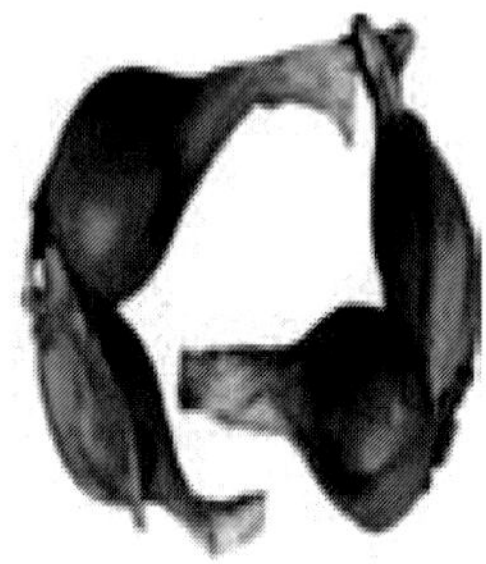

Splitting

Many factors, such as broader distances, over fertilizing, uneven watering, changes in temperatures and deep planting, can produce divisions; nevertheless, genetics are also more likely to impact their growth with some kinds divide. Moreover, larger plants tend to divide. Replication still happens, even if plants are planted closer together. It is assumed that double formation is impacted by unequal waters, particularly when the soil is dry for too long. It can also facilitate doubling by planting onions too far, as can fertilization at a pace that produces high yields. During hoeing and weeding, mechanical damage to plants initiates new growth and causes bulbs to break and double.

Management

- Stop over-fertilization and inconsistent irrigation at prescribed spacing.
- At cross-cultural activities, prevent mechanical injury.

4. Sprouting of bulbs

Sprouting

It is one of the most severe storage disorders and causes cultivators to lose a great deal. It is present in onions as well as in garlic. This condition is, however, not chronic in nature. Sprouting is more frequently recorded in white cultivars than in rose or purple cultivars (Sidhu, 2008). Excessive soil moisture at maturity and nitrogen supply are also associated with it.

Management

- Change the sowing time in such a way that it is possible to harvest in the dry season. The early harvested kharif onion and the late harvested Rabi onion do not keep well in storage.

- Retain water as soon as the bulbs reach ripeness.
- Apply a comparatively smaller nitrogen fertilizer number.
- Grow purple- or pink-coloured cultivars.
- After 75 days of transplantation, spray Chlorpropham or Isopropyl N-(3-Chlorophenyl) Carbamate (CIPC) @ 2%.

5. Poor bulb storage

Because of the sprouting, red bulbing and reduction, onions may be stored at room temperature, especially in tropical and subtropical climates, for a long time. The lifespan of onion is dependent on the season, variety, bulb dormancy, fertilizer management and irrigation, pest and disease occurrences, pre- and post-harvest management techniques and the environment for storage. Light-rooted onion such as N-2-4-1 are typically found to be able to be more stored than black, white, and yellow ones, such as Bhima Kiran, Bhima Shakti, Arka Niketan or Agri Light Red. The sprout of onion is temperature controlled. The springing rises at 10-25 °C. The relative humidity is higher rooting. In the summer (late April-May) Rabi onion is harvested in India and is treated in an environmentally friendly condition with a high temperatures and humidity. The bulbs grow quite readily as well as becoming rotten owing to fungal and bacterial conditions. For kharif onions, bulbs accumulate in winter and are used in winter and summer months and the difficulty of storage is not as significant as for Rabi crops. However, during the rabi season most onion are cultivated. Therefore, appropriate storage is needed for their extended use.

Management

- Adjustment of the planting time in order to be able to harvest in the dry season.
- 15 days before harvesting, irrigation should be halted.
- If the neck of the crop falls while the leaves are still green, harvesting should be carried out at the appropriate maturity level. The bulbs are likely to produce roots if the crop leaves on the ground until the tops are dead. Prematurely harvested thick necked bulbs do not store well. On the other hand, late harvesting leads to increased respiration, resulting in susceptibility to disease and excessive sprouting.
- Only well curated, thin necked bulbs should be selected for storage.
- If proper cleanliness is maintained in storage and the crop is sprayed after 90 days of transplantation and just before harvesting with 0.1% Carbendazim, the storage rots could be checked.

- Store the onion at a temperature of 30-35^{0}C with a RH of 65-70% under atmospheric conditions. The temperature should be held at 0-2°C in cold storage, while the RH is kept at 60-75 percent.
- For rotten, damaged and sprouted bulbs, periodic bulb turning or removal should be performed.
- For proper storage, well ventilated, enhanced storage systems with 2-3 layers of bulbs in racks or levels are necessary.

6. Greening

In the outer scales, sunlight stimulates the development of chlorophyll, which results in the scales turning green. Nitrogen applications that are excessive or late in the season can delay maturity and boost onion bulbs' greening. Greening can occur if, during the growing season, onion bulbs are exposed to sunlight or the bulbs are allowed to heal under moderate light for prolonged periods.

Greening

Management

- An early fertilisation programme that encourages foliar growth will decrease losses at bulb maturity from greening.
- Stop the application of unnecessary and late season nitrogen.
- Do not heal the bulbs in the field for long periods.

7. Sprout inhibitor injury

Ovens treated with maleic hydrazide too early in the season can be of full size at harvest, but when they are pressed, they may feel mushy and spongy. Spongitude occurs when the rings of the bulb are parted as usual and not pushed firmly to the neighbouring rings. The gaps between the rings are particularly obvious towards the collar. Maleic hydrazide inhibits germination by inhibiting the cell division. When introduced before the cells complete splitting into the bulb, the cell division stops but the cells start to expand, which causes the bulb scales to alter form and to drain from each other.

Sprout inhibitor injury

Management

- The proper timing of the use of maleic onion hydrazide is the secret for obtaining effective suppression of the sprout, while preventing damage.

- When the onion has finished generating new leaves, maleic hydrazide should be applied. When this happens, and when the bulb grows, the neck becomes hollow and breakable.
- The optimal period for maleic hydrazide is when 50 percent of plants are stuck in the field, but the plants still have 5-8 green leaves on average. The absence of at least 5 generally healthy, green leaves in the onion plant will not be sufficient to give good suppression of sprouts.
- Maleic hydrazide should be sprayed for optimal efficiency when temperatures are between 10 and 24°C and relative humidity is high.
- Low humidity can cause the leaf to crystallise maleic hydrazide and inhibit its absorption. Within 24 hours after treatment, rain will also limit absorption.

8. Sunscald

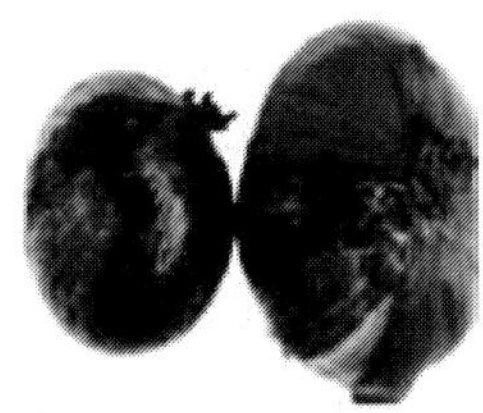

Sunscald

Sunscald is a little disease in onion plants. The temperature on the ground surface may be up to 65°C on a bright sunny day during the warm, dry springs and especially the dark organic soils. The heat damages young plants which are sensitive and kills cells on the neck. The wilts and wilts shrivel weakening tissue and strangled the neck of the vine. The amount of the injury depends on the soil surface temperature and the sensitivity of the tissue.

Management

- The only way to prevent sun scalding is to seed as early as possible before the soil temperatures get too high, so that the plants outgrow the vulnerable stage.

9. Translucent scale

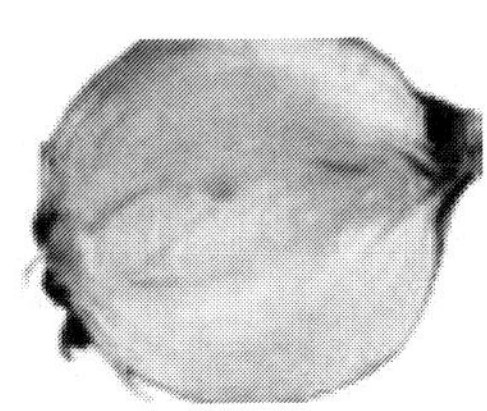

Translucent scale

On onion plants, most commonly on Spanish onions, this condition is sometimes encountered. This disease features a grey watery appearance, which is translucent on one or more layers of scales. Every layer is damaged by the illness, although there are more frequent signs of it mainly on the second and third fleshy scales. The scales involved are brownish across. The reason for the transparent size is not interpreted. This problem seems to be connected to storage, because it occurs upon harvest and after 3 or 4 months of storage worsens. Onions kept at 5 to 10°C are more vulnerable to this illness during a few weeks before ultimate storage at 0°C. In this condition, too, excessive relative humidity might induce onion.

Management

- Growers can heat-cure onion bulbs well, and then store them at 0°C and 65 to 70 percent relative humidity.

10. Wind, hail, pelting rain injury

Pelting rain injury

The seedling allium crops are highly sensitive to wind damage. Young leaves may be dried by strong, dry winds or by the drifting dirt, especially in open field areas. In a strong storm, leaves can be damaged by raindrops, which can produce irregularly white patches. These patches can grow fast and usually within 24 hours after precipitation on the side of the leaf. Depending upon the magnitude and severity of the storm, the impact of hailstones can form comparable areas or can cut the sheets or shave them. The wind and hail may cause serious damage to allium plants and must be replanted. Pelting rain does not generally produce sufficient harm in order to limit production, although spots can be confused with botrytis leaf blight, or herbicide injury. The damage can also make plants more prone to fungal diseases, by enabling nutrients to escape on the surface of the leaf.

Management

- The growers should plant cereals, like barley, on the same day that the onions are sown to shield the seedling ointments from the wind damage. Barley may be seeded in rows at 40 barley seeds per row per metre (one row of barley for every four to eight rows of onions). The radiation of cement seed at a rate of 50 to 80 kg/ha provides even better protection. A specific herbicide may be applied when the onion plants are growing to kill the barley.
- The timing of the application of herbicides is important. It should be sprayed when the seeds, if the barley is planted in rows, are around 15 cm tall. When seeded for broadcast, when it is 10 to 14 cm tall, the barley should be destroyed.
- A broad-spectrum fungicide can help protect wounded leaves from fungal infections from subsequent infection.

B. GARLIC

1. Waxy Breakdown

Waxy breakdown is a physiological disease and is thought to be more prevalent, especially close to harvest, in garlic exposed to high temperatures while growing.

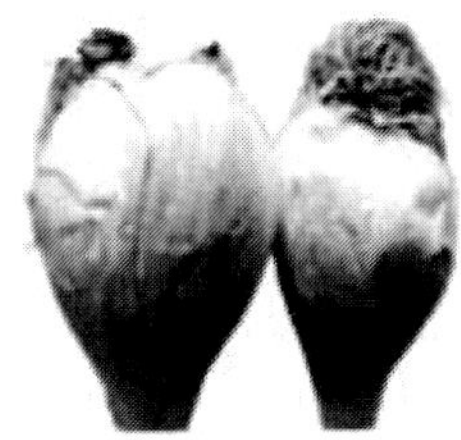

Low oxygen and poor ventilation can also lead to the creation of this defect while in storage.

In the flesh of the clove, the early signs of Waxy Breakdown appear as light-yellow patches. In storage, this progresses to the point where the clove shows a deep yellow or amber colour throughout and the clove is very transparent and sticky to the touch or waxy. The outer dry protective scales of the clove are not compromised and sometimes there is no outward sign of waxy breakdown until the advanced stage is reached where the clove shrinks, and the amber brown colour will become apparent through the clove's covering scales.

2. Tip-burn

Tip-burn

Onion tip-burn, primarily at its tips or oldest portion, may be described as a physiological disorder affecting onion leaves. It is characterized by the dying back of the very tips of the oldest and largest leaves, where the leaf progresses downwards and the whole leaf may eventually die. The leaf tips turn from green to yellow to white in the early stages as the chlorophyll disappears and the tissues die and the color changes to a light brown within a short time. This sequence progresses down the leaves and thus renders the affected portion of the leaf ineffective as a photosynthetic organ. When the onion ripens or matures in the field, this condition can be considered natural because the oldest portions of the leaves die first in this latter phase. Typically, it emerges when the onion's neck weakens and causes the top to fall to the bottom. The condition is not unnatural in this regard, but it induces a reduction in yield when it occurs early in the life of the plant. Before the plant begins to grow or mature and therefore changes with the season and the cultural climate, it does not appear at all.

Management

- The trigger is thought to be something that puts stress on the plant and may involve an imbalance of heat, drought or fertilizer.
- Balanced fertilization and persistent moisture levels help to minimize tension.
- Tip-burn is usually caused by some cultivars, and if the issue is severe, it should be avoided.

3. Ozone injury

The damage is also of interest because the leaves can be predisposed to the leaves of botrylite and purple blotch, but ozone alone causes more necrosis of the leaves and decreases bulb size. The earliest ozone damage occurs as minute white spots, particularly between the veins, in the onion's leaves. When there are vast numbers of bones, white patches of various sizes and forms with diffusive borders might merge to form. The dots are not inserted into your leaf, even though it is feasible to place larger spots. The damage caused to ozone can be characterized by the loss of defined edges and silver halos from the symptoms of botrytis leaf blight. Both ozone and botrytis lesions may also occur on the plant at the same time. The leaves become blue green to yellowish overall, when the ozone harm is severe, and die from their tips. Damages can usually be noticed one to three days after a rainy, foggy and moist occurrence happens when air pollution is significant. Big occurrences are referred to as ozone levels and generally produce spoils on onion leaves of 15 parts per cent million in ambient air. The youngest and the older leaves are often more ozone resistant.

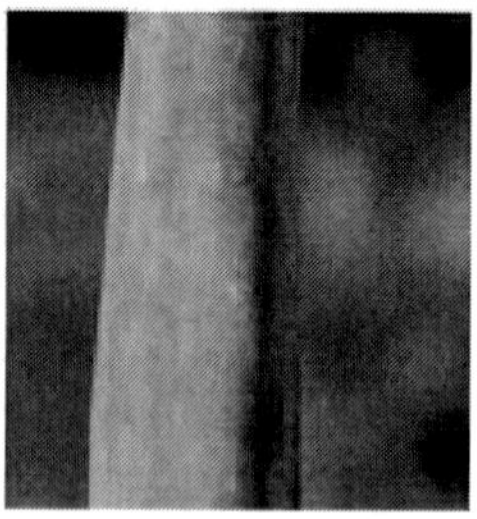

Ozone injury

Management

- Other than identifying onion fields in areas with low air pollution, there are no control measures available.

4. Herbicide injury

Herbicides may cause allium plants temporarily or permanently. Damage can be caused if these products are applied at excessive rates, in poor weather or at a wrong development phase. The contact herbicides are used to generate brushes and necrotic patches, while leaf tips may wither with translocated herbicides. The plants usually recover and symptoms decrease over time. However, the damage would be irreparable if the application rate is too big. Damage can also be noticed if allium plants have been cultivated while a residual herbicide was taken from the previous year's crop that is sensitive to them.

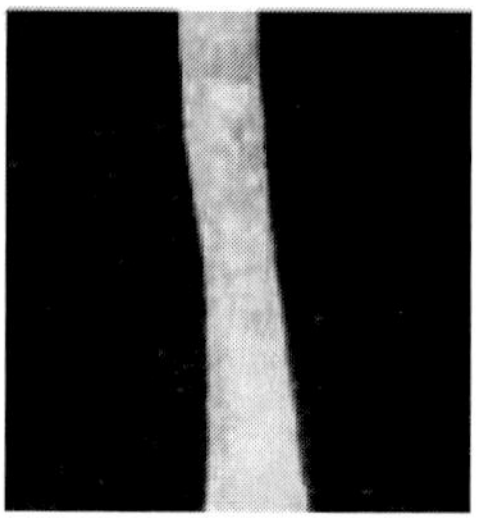

Herbicide injury

Management

- Growers should meet guidelines about rates, water volume and growth stage for proper application in order to prevent herbicide injury.
- Records of weather and crop stage conditions are useful when applying herbicides in case of damage.
- Excessive wetting of onion foliage with improper mixtures will result in serious crop injury.

6

Physiological Disorders of Root Crops

A. RADISH

1. Forking

Forking

There is secondary elongating growth in the roots that gives a form of fork like structure to the root. This occurs in heavy soil due to soil compactness. An elongated radish root is encouraged by un-decomposed organic manure. It can be fixed by decreasing field moisture, balanced irrigation and also by sowing radish and carrot in sandy loam or light soil with loose and friable soil in nature.

Management

- Varieties with a root system vigorously elongated should be picked.
- In areas of shallow arable land, short root or round root should be selected.
- In deep soil conditions, poorly elongated roots, long root and giant root varieties should be cultivated.

2. Pithiness (Pore extent)

Pithiness

One of the key determinants of the consistency of radish tubers is also known as sponginess. This pithiness is marked by the death of the cells in which gas accumulates in xylem parenchyma. Pithy tissue originates from the largest cells that are comparatively far from vascular tissue strands and situated halfway between the middle and the tuber periderm, the elongation of the tuber, the dying cells can be broken apart, leading gradually to hollow tubers.

Management

- Harvesting should be carried out at a reasonable time.
- By crossing between varieties with high assimilation ability and early growth, varieties with early root elongation and late pore growth could be created.

3. Hollow root

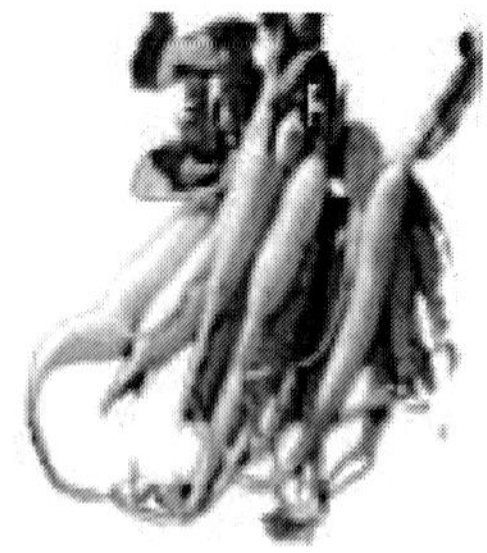

It prevents the formation in the centre of roots of secondary meristems, leading to the creation of inter-cellular spaces that would otherwise have been filed with parenchyma. It is due to high temperature during the root elongation period 16-30 days of sowing.

4. Akashin

Akashin did not occur until the tap root reached the vigorous growth stage (when it weighed about 300 g) and then developed with the growth of the tap root. The root growth is tested. Generally, this condition arises because of boron deficiency. It is caused by day and night temperature variations. Akashin occurred particularly on radishes sown from early June to mid-July.

Management

- Spray 1-2 ppm of boron to rectify this disorder.

5. Bolting

Bolting

Bolting is referred to as seed stalk production without proper development of the economically essential / edible vegetative section. Radish is a crop that vernalizes seeds in response to low temperatures. Because of its disadvantage in cultivation, the degree of bolting capability has been studied and breeding research has been geared towards late bolting.

Management

- Sowing of late bolting varieties.
- Early bolting plant removal from the late spring and summer varieties.

6. Pungency

One of the most important quality traits is the flavour of radish for fresh eating. 4-methylthio-3-butenyl isothiocynate is a potent part of radish. Many (temperate) varieties have been reported to have a low 4-methylthio-3-butenyl glucosinolate content (100-200 mmol per 100 g fresh weight). Owing to high temperatures and water stress conditions, this occurs.

Management

- Proper control of water.
- Variety collection as per season

B. CARROT

1. Root splitting

Root splitting

In some carrot growing areas, splitting or cracking of carrot roots is a major problem. This tendency is believed to be regulated by genetic factors, but there are also a number of other factors involved. The roots are destroyed, as its name suggests, making it unfit for the market. Earlier cultivars break up quicker than late cultivars. Root splitting is increased in soils with a high content of nitrogen. The high soil ammonium compound concentrations produced more splitting than other sources of nitrogen. Strong side dressings with NH_4NO_3 increased splitting when the roots were 6.5 cm in diameter, especially when nitrogen was applied to widely spaced roots. If roots are broad, as opposed to small ones, there are chances of division.

Management

- Grow resistant varieties.
- Sow the seeds at close spacing.
- Harvest the crop at right maturity stage.
- Follow proper management methods, such as correct spacing and optimal N application quantity.
- Prefer to N fertilizer source types other than ammonium compounds.
- The soil should be free of rocks, as they will prevent the roots from being real, twisting them as they grow.
- A lot of moisture (1 inch of water per week) is needed to help the carrots grow quickly, but to stop carrots cracking.

2. Cavity spot

Cavity spot

This disorder arises in the cortex with the subtending epidermis that is collapsing in most cases from a pitted lesion. Ca deficiency is triggered by the condition of the cavity spot. This is correlated with an increased accumulation of K, which results in a decreased accumulation of Ca. Increasing the amount of Ca in the increasing medium results in a significant decrease in the occurrence of cavity spots.

Management

- Increasing the amount of Ca in the rising medium results in a substantial reduction in cavity spot incidence.
- Application of fertilizer should be held to a minimum and should only be carried out in spring or early summer when the crop requires it most. This will avoid the accumulation of K in the plant.

C. TURNIP AND BEET ROOT

1. Internal browning

In the middle of the root, this disturbance appears brown. This condition typically develops during the root ripening phase and high temperature owing to heat stress. The level of internal bruneness is directly linked with the fluctuations in the polyphenol biosynthesis enzyme activity and the glutathione cycle ascorbate.

Internal browning

Management

- It can be controlled by selection of varieties according to their adaptability and seasons.

2. Internal black spot or brown heart

Brown heart

Boron deficient plants usually remain dwarf or stunted. The greenery is smaller than average. The roots do not grow to full size and, under conditions of severe boron deficiency, remain very small and twisted and have a rough, unhealthy, grayish appearance instead of being clean and smooth. They also have wrinkled and broken surfaces. Strong or corky spots inside the fleshy roots are located scattered around the roots. Discoloration of cell walls and granulation of meristematic cell protoplasts occur first. In the cambium, with little differentiation into xylem and phloem, cell division continues and many cells are abnormally large. This results in disrupted translocation, and necrotic black spots are eventually followed by distorted development.

Management

- The amount of borax necessary for satisfactory control of boron deficiency varies with the nature of the soil, the response of the soil and the moisture of the soil.
- It is recommended to add 10-15 kg of borax per hectare.

3. Zoning

Zoning

The beet root is hardy and likes a cool atmosphere at low temperatures. While it develops in warm weather, the best in cool weather is the growth of colour, texture, sugar content etc. of roots. High temperature induces zoning, i.e. alternating light and dark red concentric rings in the root appearance.

Management

- It can be controlled by selection of varieties according to their adaptability and seasons.

D. CELERY

1. Black heart

Black heart

Blackheart is a celery physiological reaction associated with inadequate assimilation of calcium that occurs in most regions of celery production and is closely related to growing conditions. The entire crown could be lost in a few days if conditions conducive to the creation of the blackheart continue. The plants may recover or appear to recover in minor cases, while symptoms may return. The occurrence of symptoms is much more serious as plants reach maturity. The presence of symptoms is closely related to fluctuating levels of water and the calcium available. Environmental conditions that favour rapid growth, such as heavy rain or irrigation, facilitate the development of symptoms as a result of drought. Also, high levels of nitrogen, potassium and sodium can play a role. Some cultivars of celery are more vulnerable to black hearts.

Management

- By maintaining steady plant growth, Blackheart is avoided, preventing large variations in moisture and nutrients.
- Soluble calcium drenching applications can reduce or prevent blackheart growth.
- Plant cultivars of celery which are not susceptible to the syndrome.
- Drip irrigation that provides even higher levels of moisture will help reduce the risk.

2. Cracked Stem (Brown Checking)

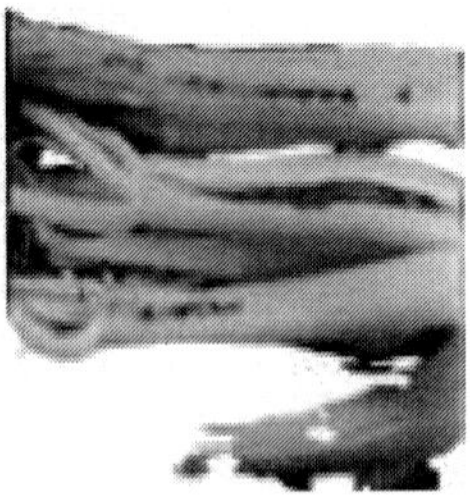

Cracked Stem

Over the whole mature petiole, a transverse cracking of the epidermis rarely occurs. The epidermal layers turn brownish black, and the tissue turns brown directly below them. Leaf mottling, root browning and dying, brittleness and some brown, shallow, longitudinal streaks of the fleshy petiole may accompany this characteristic sign.

Management

- Careful management of fertilization and irrigation.
- It can be corrected by the application of calcium.

3. Pencil strip

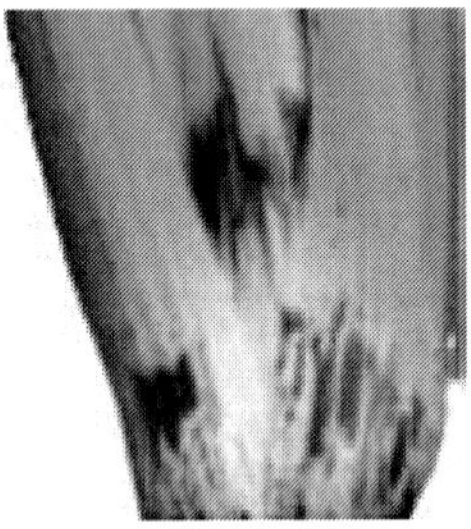

Pencil strip

A disorder associated with high soil phosphorus showing narrow brown lines on petioles. Optimum application of phosphatic fertilizers helps to minimize this condition.

7

Physiological Disorders of Leafy Vegetables

A. LETTUCE

1. Tip burn

Tip burn

Tip-burn is the marginal collapse and necrosis of rapidly spreading inner lettuce leaves, at or near the leaf margins. Typically, the condition occurs near harvest, when it may lead to complete crop failure. Early signs include discoloration of the veins and/or the growth of dark brown to black spots near or at the margins of the leaves. The necrotic areas coalesce as the disease progresses, creating a lesion up to several centimeters long. Tip-burn is thought to be a calcium-related condition and is associated with low calcium tissue concentration, which is critical for the strength of the cell wall and the integrity of the membrane. Due to the low mobility of calcium ions inside the plant, warm temperatures, excessive fertilization, an increase in light intensity and other factors that lead to the rapid growth of lettuce may boost the production of tip burn.

Management

- Harvesting 3 days earlier in well-fertilized soils can minimize tip-burn by as much as 30 to 50 percent.
- To optimize water and nutrient absorption, make sure the crop is supplied with enough water overnight or early morning.
- In summer plantings, consider using tip-burn-tolerant cultivars.
- With crops planted in mid-summer, better management practices are more important.

2. Russet spotting

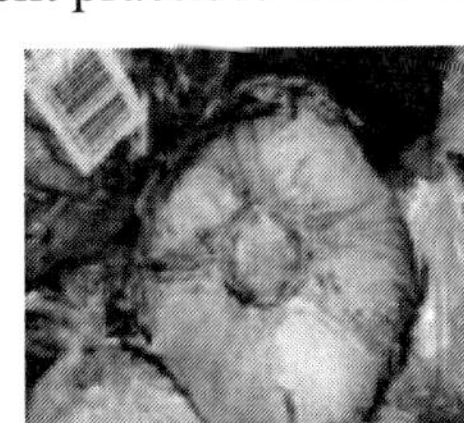

Russet spotting

Russet spotting is a physiological disorder that stimulates the development of phenolic compounds which lead to brown pigments due to exposure to high ethylene concentrations. Russet spots occur particularly on the midribs of leaves as small dark brown spots, which increase in number with storage periods.

Management

- Maintaining temperatures close to 0 ° C, keeping in a low O2 environment and not storing goods producing ethylene in the same room decreases the production of russet spotting.

3. Brown Rib

This disorder is characterized by yellowing and tanning of outer leaves, which usually appears, but not exclusively in storage and is aggravated by the presence of ethylene (C_2H_4). It is aggravated by high temperatures and humidity especially after it has rained just before harvest.

Brown Rib

Management

- Planting head lettuce on raised beds and avoiding irrigation during the last two weeks before harvest will reduce the incidence of brown rib.
- This issue can be alleviated by the use of resistant cultivars.

4. Puffy Heads/ Ribbiness

Symptoms of puffy heads are loose, soft heads or ribby, which prevents normal head development or results in rough, loose heads. Prominent outer leaf ribs which do not fit around the head are characteristic of disorder. This disorder is caused by among other factors high temperature, rain and high humidity near harvest. Shading by wind breaks, cloudy weather and low soil fertility also seem to increase the frequency of the disorder, which suggests that it may be caused by low carbohydrate synthesis in the plant.

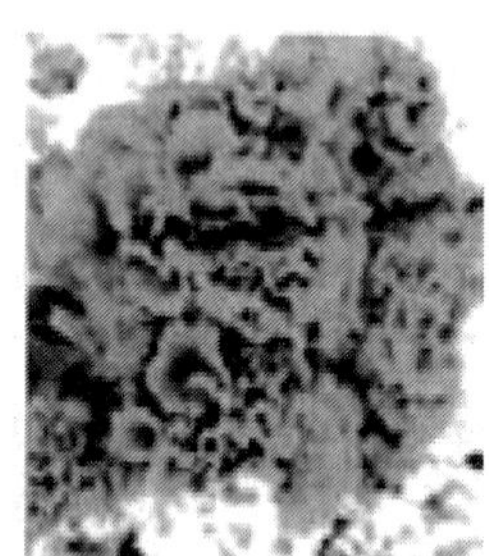

Ribbiness

Management

- To overcome this disorder circumstances which lead to low carbohydrate synthesis should be avoided *eg* shading.
- One of the ways to control this disorder is by use of adapted cultivars.

5. Rib discoloration

Rib discoloration

This disorder is characterized by light brown to black; oblong areas discolour the inner surface of midribs. This disorder is caused by high temperatures during growth.

Management

- There is no control of rib-discoloration but irrigating during hot weather can help to reduce it.

8

Physiological Disorders of Cucurbitaceous Vegetables

A. CUCUMBER

1. Blossom end rot (BER)

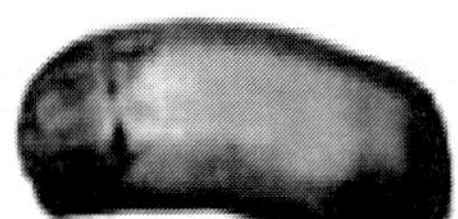

Blossom end rot

In several cucurbits as well as other crops, BER is a physiological condition found (for example, tomato, brinjal and pepper). It typically occurs as a general rot at the flowering end of rising fruit. Usually, the blossom end rot is the result of inadequate or uneven irrigation, high humidity, or other variables that slow the flow of water through the plant. Slow water movement may also result in temporary calcium deficiencies leading to blossom end rot as calcium is added into the plant with the transpiration stream.

Management

- Provide ample fertility and adequate calcium irrigation.
- Do not use elevated levels of ammonia fertilizer, which may make this issue worse.
- Stop damage to roots.

2. Drought stress

Drought stress

Cucurbits are especially susceptible to drought. Usually, fruits are 85 to 90 percent water and can suffer under conditions of drought. Pumpkins also grow long vines with many leaves and, during hot summer days, large amounts of water will transpire. Extreme drought stress influences the production of fruit, contributing to unmarketable produce. At the end of the blossom, the affected cucumber fruit may appear twisted, distorted or tapered; pumpkins become soft and wrinkled. Furthermore, pumpkins stressed by drought struggle to gain adequate size, affecting yields. A lack of foliage during drought can also contribute to the fruit being sunburned.

Management

- Irrigate when appropriate.

3. Flood damage

Flood damage

Nutrient deficiencies or generalized yellowing sometimes occur as symptoms. Anaerobic (low oxygen) conditions for plant roots can result in prolonged exposure to flooded soils, ultimately causing death. The plant is also unable to take up enough nutrients when large numbers of roots die, leading to nutrient deficiencies.

Management

- Although flood damage is always inevitable, planting in elevated beds will enhance drainage.

4. Poor pollination

Poor pollination

Cucurbits require pollination for the production of fruit, with the exception of parthenocarpic cucumbers. To ensure adequate fruit growth, several visits from pollinators on the day that a flower is open are often needed. When pollination is low, many fruits may appear misshapen and thin. Cucumbers at the end of the fruit stem will be reduced in size. Very high and low temperatures, resulting in poor pollination, may also impact pollen viability. It may be difficult for bees to locate the flowers if too much nitrogen is used (resulting in excessive vegetative growth) or the plants have been spaced incorrectly.

Management

- Supply pollinators to make sure high yields and a healthy fruit collection.
- Do not spray insecticides when flowers are open and during the morning hours, insects are actively pollinating plants.

5. Light Belly Color

The undersurface of the cucumber fruit stays light in colour instead of turning dark green, characterizing this condition. Fruit lying on cool, moist soil is commonly found.

Management

- By preventing luxuriant vine growth, it can be partially regulated. Stop nitrogen in abundance.

6. Oedema

This is a physiological condition that occurs more rapidly as roots suck up water than the plant can use or transpire via the leaves. In the internal cells, water pressure builds up, causing them to burst, leaving dead cells visible as blisters.

Oedema

7. Pillow

Owing to the low amount of calcium in the tissue, it is a fruit disease of the processing cucumber. In this disorder, in the mesocarp of the newly harvested fruits, an irregular white styrofoam such as porous textured tissue is produced. Vascular tissue can collapse and become necrotic in some pillow areas.

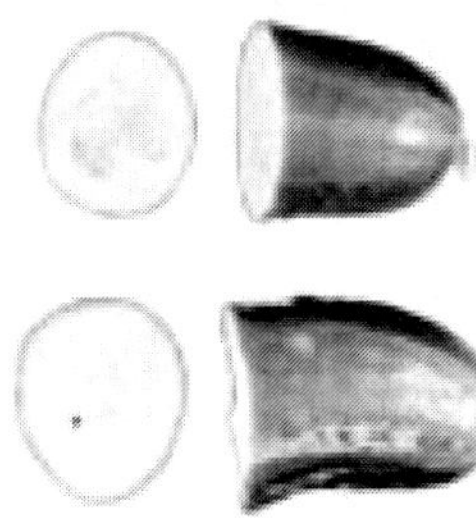

Pillow

8. Leaf Silvering

It is physiological disorder of summer squash (*Cucurbita pepo*). Owing to moisture shortage, this disorder happens. The leaves are coloured silver and contain less chlorophyll. Silver leaves have paces between and between the mesophyll cells of the upper epidermis and the mesophyll cells. With less spongy mesophyll cells, the palisade cells are identical. Photosynthesis in the silver leaves is hindered.

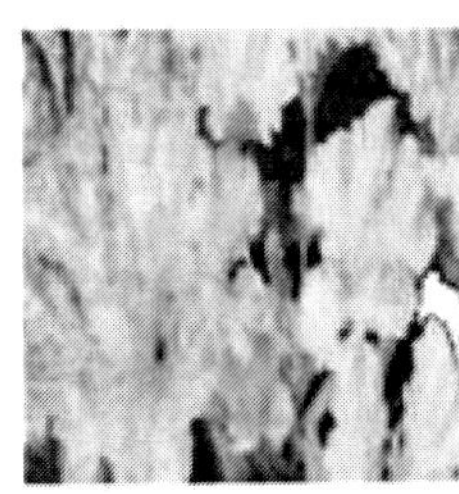

Leaf Silvering

Management

- Irrigate when necessary

B. WATER MELON

1. Hollow heart

The Hollow Heart is the creation of a hollow cavity within some cucurbit fruit. A number of variables can contribute to this condition, including low boron levels, genetics, and uneven water regulation. The hollow heart, while not visible from the outside, makes fruit unmarketable.

Hollow heart

Management

- Ignore cultivars that appear to represent hollow cores.
- Make sure the adequacy of levels of boron in soil.
- Be cautious not to fertilize too much.
- Follow the prescribed spacing of plants and prevent irregular irrigation.

2. Rind Necrosis

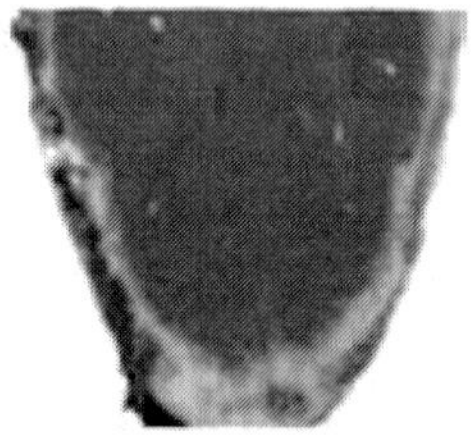

Rind necrosis

In fruit rinds, dead, hard, dry, reddish-brown to brown spots or tissue patches typically occur in either cantaloupe or watermelon. The affected areas range from 3 mm (1/8") spots around the entire rind to major dead areas in size. Symptoms from outside watermelons are not visible and are rarely found in the flesh. In the cantaloupe, dead tissue may reach into the flesh of the fruit. On the cantaloupe fruit surface, circular, water-soaked depressions often form. It is suspected that the onset of this disease can be triggered by environmental factors that bring stress on the plants. Among the forms, rind necrosis susceptibility varies. The disease occurs sporadically and is suspected to be associated with bacteria that may be present in fruit, although the reasons for the development of symptoms are not clarified. Melons are also known to be predisposed to drought pressure.

Management

- Genetic resistance in watermelons has been established. Stop dry melon stress.

3. Delay in fruit ripening

Delay in ripening

For muskmelon and watermelon, this problem is especially important. Ripening delay is often associated with less sweetness and fruit cracking due to high moisture levels and variations in temperature at the ripening point.

Management

- At the last stage, irrigation should be stopped to hasten ripening.
- Sowing time should be changed in such a way that fruits are ripe in hot and rainless conditions, which speeds up ripening and increases fruit flavour at the same time.

4. Stem splitting

Stem splitting

Stem splitting is most frequently found in transplant production when temperatures are low or when there is a period of rapid growth resulting from elevated temperatures, increased irrigation or high fertility. In mild cases, plants with little adverse effects may be transplanted, but in severe cases, seedlings should not be transplanted if possible.

Management

- Provide seedlings with warm, uniform temperatures and allow for even growth during the development of transplants.

C. MUSK MELON

1. Chilling injury or cold injury

Chilling injury

Low temperatures on the vine (cold injury) or in post-harvest storage can damage the greenhouse cucumber fruit (chilling injury). By maturing fruit that is subject to cold draughts and sudden drops in temperature, characteristic scars are produced. These consist of white or light brown longitudinal marks, often on one side, but often all around the fruit. In appearance, the wounds are corky and, though superficial, they detract from market quality. Thrips that feed on the developing fruit may result in similar symptoms. The fruit may split in severe cases. It could be bitter for a slow growing fruit.

Chilling injury, due to the collapse of epidermal tissues, contributes to fruit pitting. These pits turn yellow and can eventually cause the entire fruit to become chlorotic. By chilling damage, the susceptibility of fruit to microbial decay can be increased.

Management

- The risk of cold injury can be reduced by minimizing cold draughts in the greenhouse and avoiding spraying plants with cold water.
- Greenhouse air temperatures should not be permitted to fall below 16°C.
- Chilling injuries can be avoided by storing cucumber fruit between 10 and 13°C.

2. Measles

Smooth-skinned melons and cucumbers have the most noticeable symptoms. There are tiny brown patches scattered across the fruit's surface. The spots are on the surface and do not move outside the fruit's outer epidermal layers. On leaves and branches, these spots can also occur. In connection with ecological conditions that encourage guttation. The guttate droplets contain high salt concentrations that burn the epidermis. Measles spots occur where a droplet of guttation has formed.

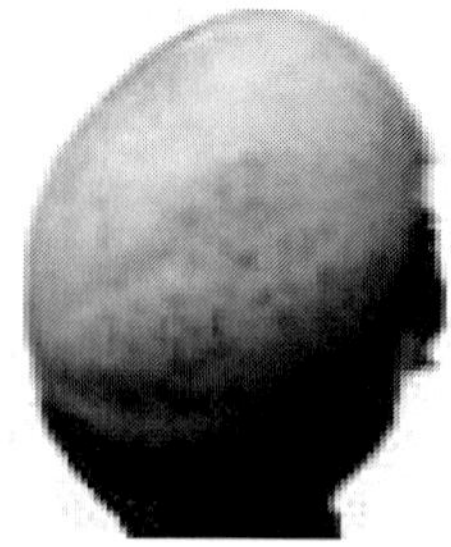

Measles

Management

- Monitor measles by lowering the frequency and length of irrigation in fall-harvested crops as fruit reach maturity.
- Irrigation reduction has not shown any detrimental effects on fruit size and soluble solid content at the later stages of fruit production.

3. Sunscald

On fruit, papery white areas grow. It grows when fruits are unexpectedly exposed to direct sunlight during hot summer weather.

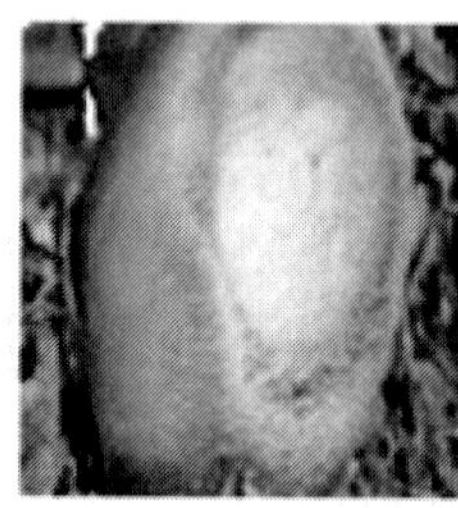

Sunscald

Management

- Minimize to ensure the fruit is protected by maintaining high vine growth.

D. POINTED GOURD

1. Unfruitfulness

A dioecious cucurbit is a pointed gourd, and staminate and pistillate flowers are produced in separate plants in the sex type. So, there are different male and female plants. The fruit is produced by female plants, while male plants serve as pollen donors. In order to ensure adequate pollination, fertilization and fruit set-up, the appropriate number of male plants should therefore be present in the female plant population. Where pistillate flowers drop off in female plants due to lack of pollination and fertilization, a common problem is discussed. In some cases, due to parthenocarpic stimulation that also abscises after a few days, the ovary of the unfertilized flower may expand a bit.

Unfruitfulness

Management

- In order to ensure adequate pollination and fruit setting, male plants must be grown in the field along with female plants at a ratio of 10-12 male plants per 100 female plants.
- To achieve an adequate fruit collection, hand pollination can be achieved effectively.
- Plant protection chemicals should not be applied to the plants in the morning hours as it hinders the movement of pollinizer insects.

9

Physiological Disorders of Okra and Legume Vegetables

A. OKRA

1. Poor seed germination

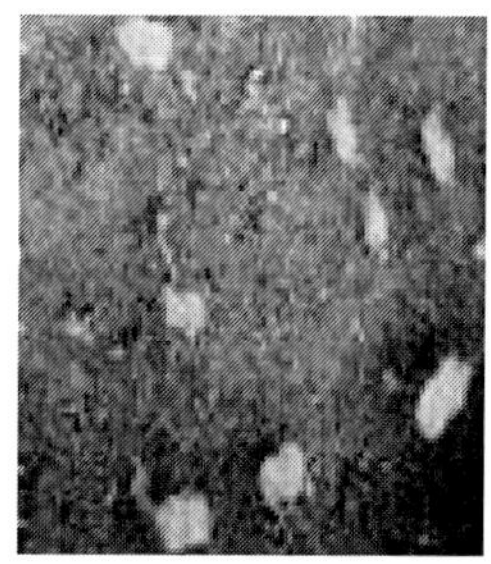

Poor germination

When the soil temperature stays at or below 20 ^{0}C, seeds germinate poorly. During early spring-summer planting, this issue of seed germination occurs when seeds are to be sown at low temperature conditions.

Management

- 24 hours of soaking the seeds in water.
- Soaking the seeds in hot water for half an hour at 45°C.
- Alcohol seed therapy for half an hour.

2. Chilling Injury

Discoloration, pitting, water-soaked lesions and increased decay are common signs of chilling okra injury (especially after removal to warmer temperatures, as during marketing). In terms of their susceptibility to chilling injury, different cultivars may differ. In order to minimize chilling symptoms, calcium dips and modified atmospheres have been recorded.

B. COWPEA

1. Sunburn

The veins were reddish-brown in color and some of the interveinal leaf tissue was reddish-brown to tan in color. This reddish-brown discoloration of veins and interveinal tissue on the lower surface of a portion of a leaflet appears to be sunburn.

Management

- Irrigation at regular interval

- Planting wind breaks across the field offers protection against desiccating hot waves, resulting in low sunburn.
- During hot hours, irrigation via the sprinkler system increases moisture, cools the field atmosphere and thereby reduces the occurrence of sunburn.
- Only light irrigation with increased frequency (4-5-day interval) is considered beneficial in light and sandy soil.

MCQ'S ON MAJOR PHYSIOLOGICAL DISORDERS OF VEGETABLE CROPS

1. BER is a physical disorder of

   a) Cucumber b) Tomato
   c) Brinjal d) All of them

2. Golden flake is a physiological disorder of

   a) Brinjal b) Potato
   c) Okra d) Tomato

3. Browning in cauliflower is caused due to deficiency of

   a) Calcium b) Boron
   c) Zinc d) All of them

4. Cat face is a physiological disorder of which vegetable crop

   a) Potato b) Tomato
   c) Brinjal d) Okra

5. Blossom end rot is a very common problem on

   a) Green tomato b) Ripe tomato
   c) Over ripe tomato d) a and b both

6. Blossom end rot is caused by deficiency of

   a) Calcium b) Boron
   c) Zinc d) All of them

7. Types of cracking may develop on tomato fruit

   a) Radial cracking b) Concentric cracking
   c) Cuticular cracking d) All of them

8. Fruit cracking is caused by deficiency of

   a) Calcium b) Boron
   c) Zinc d) All of them

9. Catfacing is most common in which type of tomato fruits
   a) Small fruit b) Medium fruit
   c) Large fruit d) All of them
10. Sunscald occurs on green tomato fruit is due to
   a) Low temperatures b) Exposed to sun
   c) High temperature d) b and c both
11. Brown heart of turnip is cause by the deficiency of
   a) Boron b) Water stress
   c) Calcium d) High temperature
12. Splitting in carrot is basically due to
   a) Boron b) Water stress
   c) high Nitrogen d) High temperature
13. Pillow is a physiological disorder of
   a) Cucumber b) Radish
   c) Carrot d) Beet root
14. Frog eye rot is a physiological disorder of
   a) Potato b) Chilli
   c) Carrot d) Tomato
15. Growth crack is a physiological disorder of sweet potato due to
   a) Moisture imbalance b) Water stress
   c) Calcium d) High temperature
16. 'Russeting' is a physiological disorder of
   a) Chilli b) Potato
   c) Carrot d) Tomato
17. The blister like swellings on the leaf resemble an undifferentiated callus type growth is
   a) Russeting b) Edcma
   c) Herbicide injury d) Puffiness
18. Series of dark green spots arranged in a line from the stem end to the blossom end is known as
   a) Russeting b) Zippering
   c) Zebra Stripe d) Calyx withering

19. Foliar application of 50 ppm NAA at full bloom stage of chilli effectively controls
    a) Flower drop b) Fruit drop
    c) Poor fruit set d) a and b both
20. White heart is a physiological disorder of
    a) Carrot b) Watermelon
    c) Muskmelon d) Beet root
21. Whiptail disorder of cauliflower is caused due to deficiency of
    a) Molybdenum b) Boron
    c) Zinc d) Calcium
22. The development of small premature curds while the plants are young is known as
    a) Whiptail b) Riceyness
    c) Buttoning d) Hollow stem
23. Blindness disorder occurs in
    a) Cauliflower b) Cabbage
    c) Brussels sprouts d) b and c both
24. Premature initiation of floral buds on the upper surface of curd is called as
    a) Fuzziness b) Riceyness
    c) Buttoning d) Hollow stem
25. Hollow stem is a physiological disorder of
    a) Cauliflower b) Cabbage
    c) Brussels sprouts d) b and c both
26. Premature seed stalk development is called as
    a) Blindness b) Bursting
    c) Bolting d) Hollow stem
27. Heads split or burst of cabbage is usually due to
    a) Heavy rains b) Overwatering
    c) Delayed harvest d) All of them
28. Splitting and doubling of bulbs is caused by
    a) Wider spacing b) Uneven watering
    c) Over fertilization d) All of them

29. Sprouting in onion is more common in which cultivars
    a) White cultivars b) Pink cultivars
    c) Purple cultivars d) None of them
30. Waxy breakdown is a physiological disorder of
    a) Onion b) Garlic
    c) Tomato d) Watermelon
31. The cavity spot disorder of carrot is induced by deficiency of
    a) Calcium b) Boron
    c) Zinc d) Molybdenum
32. Appearance of alternate light and dark red concentric rings in the root of Beet root is known as
    a) Black heart b) Zoning
    c) Brown heart d) Cracking
33. Cracked Stem is a physiological disorder of
    a) Carrot b) Lettuce
    c) Turnip d) Celery
34. Maleic hydrazide inhibits
    a) Bolting b) Browning
    c) Sprouting d) Greening
35. Secondary elongating growth in the roots that gives a look of fork like structure is known as
    a) Forking b) Pithiness
    c) Hollow root d) Bolting
36. Akashin disorder of radish is generally occurs due to deficiency of
    a) Calcium b) Boron
    c) Zinc d) All of them
37. Leaf Silvering is physiological disorder of
    a) Summer squash b) Cucumber
    c) Winter squash d) Water melon
38. Stem splitting of water melon is due to
    a) High temperature b) High fertility
    c) Increased irrigation d) All of them

39. Fruit having thin scars that extend partially or fully from the stem scar area to the blossom end is known as

    a) Zebra stripes b) Zippering

    c) Vascular browning d) All of them

40. Internal rust spot or internal brown spot symptoms are quite common in potatoes grown in soils

    a) Sandy b) Sandy loam

    c) Clay loam d) Silty

Answers

1	d	**9**	c	**17**	b	**25**	a	**33**	d
2	d	**10**	b	**18**	c	**26**	c	**34**	c
3	b	**11**	a	**19**	d	**27**	d	**35**	a
4	b	**12**	c	**20**	b	**28**	d	**36**	b
5	d	**13**	a	**21**	a	**29**	a	**37**	a
6	a	**14**	b	**22**	c	**30**	b	**38**	d
7	d	**15**	a	**23**	d	**31**	a	**39**	b
8	b	**16**	d	**24**	b	**32**	b	**40**	a

Section 3
Major Physiological Disorders of Fruit Crops

10

Physiological Disorders of Mango

1. Alternate Bearing

This is one of the most serious difficulties in mango farming since it makes mango agriculture less rewarding for farmers. In one year (a year) and very little or no harvest next year, mango trees appear to bear huge crops (off year). This bieniality is therefore also present in normal ranges (Singh, 1990). The majority of commercial species in north India include Dashehari, Langra and Chausa, but the South Indians are common in kind, such as Totapuri Red Tiny, Bangalor and Neelum, but they are unable to yield excellent fruits under north Indian circumstances. In general, when loaded with berries mango trees do not develop new shoots. If new shootings are generated even after harvesting, the number is little and does not bloom in next year because fresh vegetative flush of mango required a specific amount of maturity (8-10 months) for the differentiation of floral buds. In general, if flowers and fruit are generated this year, new shoots will be formed in the next months of March-April that blossom again next year to provide alternative flora and fruit. In addition, if a tree cultivates heavy plants during one season, it will become exhausted and not produced in the next season. The ratio of C/N (carbon/nitrogen) and hormone imbalances also play a key part in the causation of mango bearing changes, in addition to environmental factors. The problem is caused by genes, physiology, the environment, diet and hormones (Bhargava *et al.*, 2011).

Management

- The outcomes in early vegetative development and start to blossom in September are paclobutrazole (PP333) @ 4 grammes / trees as soil and foliar spray.
- Unloading of some 'on-year' flowers and tree pruning immediately after harvest will provide an adequate physiological balance between vegetative and reproductive growth and will allow sufficient solar light to reach the inside of the orchard, thereby improving the yearly tree production.

- In addition, pruning helps to develop new branches that can grow quickly after harvest in the following flowering season. Flowering can also result from 'off' year smudging Mangoes are formed when slow fires are created that generate smoke under the mango tree. The use of Ethrel may potentially lead to flora.
- Cultivars such as Amrapali, Mallika, Ratna, Dashehari-51, Pusa Arunima, Pusa Shrestha, Pusa Pitambar, Pusa Lalima, Pusa Pratibha, Arka Anmol, Arka Aruna, Arka Puneet, Arka Neelkiran can be selected for commercial cultivation by growers in the primitive stage of mango cultivation, as they are frequent in nature. It is also recommended that the old mango tree be rejuvenated by grafting or budding with regular bearing varieties to turn the alternative bearing habit into a regular one.

2. Mango Malformation

Farmers, particularly in northern India, have faced huge problems due to this disorder over the last few decades. Although it was first noted in Bihar in 1891, it is now an alarming hazard in Punjab, Delhi and Uttar Pradesh, with Gujarat, Maharashtra, Bihar, West Bengal and Orissa to some extent (Prasad *et al.*, 1965). Southern India is basically free of this threat. This condition was documented by Burns and Prayag (1920). There are about 50-80 percent casualties due to this condition (Singh and Chakravarti, 1935). It consists primarily of two forms: vegetative and floral. In nursery seedlings and in young plants, vegetative malformation is more common (Nirvan, 1953). Malformed vegetative shoots with thin scaly leaves assume the appearance of bunchy tops. In clusters, vegetative buds expand, swell and lose apical dominance.

Malformation

The most damaging one, however, is the floral malformation that occurs at the bearing stage on the trees and creates small, thicker and hypertrophied panicles as condensed clusters. The proportion of male and hermaphrodite sterile flowers increases in deformed panicles and no fruit is seen in those branches. Flower Mango Mishap Most commercial mango varieties such as Bombay Orange, Dashehari, Lucknow Safeda and Chausa are extremely prone to the illness, although they are less impacted by Baramasi. Cultivars like Ellaichi, Alib and Bhadauran, however, are not of outstanding fruit quality and are used solely for commercial production for resistive breeding. They are free of all this illness. Scientists from various parts of the world have confirmed the *Fusarium moniliformae* var. fungi. This disease is closely correlated with subglutinans. Temperature ranges of 21-27°C (max.) and 8°C (min.) along with relative humidity of 85% are conducive to fungal growth during the late winter or early spring, resulting in a serious incidence of malformation on newly emerging panicles.

Management

- Sirohi *et al.* (2005) indicated that before fruit bud differentiation followed by debloosming of newly emerged panicles or buds during January-February, the application of NAA/ Planofix (200 ppm) during October is beneficial to regulate the disorder.
- Equally effective in minimizing floral malformation is spraying various phenolic compounds such as catechal, cynamic acid, tannic acid during the 1st week of October at 2000 ppm.
- Several anti-malformins were suggested when sprayed on panicles just after emergence for beneficial effects (4-6 cm). It was also found that glutathione sprays at 2250 ppm, ascorbic acid at 2110 ppm, and $AgNO_3$ at 600 ppm were effective for malformation control.
- It is very efficient to reduce the incidence of malformation by applying nutrients such as phosphorus & potash and micronutrients, particularly zinc & boron, just after harvesting. Moderate nitrogen application has helped to decrease vegetative malformation, while zinc and boron have led to auxin biosynthesis, which eventually helps to reduce the occurrence of mango malformation.
- Removal at a distance of 15-20 cm below the point of occurrence of just developed malformed panicles or vegetative shoots is useful to further mitigate the spread of disorder.

3. Black Tip

Woodhouse first noted this mango disease in 1908 (Zhang *et al.*, 1995). In Punjab, Uttar Pradesh, Bihar and West Bengal, this disease is very prevalent. It causes the growers substantial economic losses. Dashehari is highly vulnerable among commercial cultivars. The disease was mostly seen in gardens around brick ovens. The gases that make up the briquito kiln fumes, including as carbon dioxide, sulphur dioxide and ethylene harm a growing fruit tip and cause black tip symptoms. Ethiopia of the distal fin of fruit leaving the mesocarp and the seeds intact, accompanied by the appearance of the grey patches, which turn dark and coalesce to turn brown over the whole fruit tip. In severe situations, a thin layer of collapsed tissues remains in the injured fruit tissue. The necrotic region is frequently limited to the fruit tip due to the high stomatal density, strong catalase, and peroxidase activity.

Black tip

Management

- Preventive actions for the condition would be to build the orchards at least 1.6 km east to west away from the brick kiln and 0.8 km south to north and elevate the height of the kiln to at least 15-18 m.
- 1% borax (10g/litre) is highly efficient to spray thrice first before blooming, secondly during blooming and thirdly at fruit setting. Further more, at pea level, the acidic gases from brick kilns are precipitated on the tip of fruit by sprinkling additional alkaline solutions such as caustic soda (0.8 per cent) or washing soda (0.5 per cent) with two more spraying at 15-day intervals (Sharma, 2005a).

4. Spongy Tissue or Soft Nose

Spongy tissue is a major physiological disorder of 'Alphonso' mango which causing about 30% fruits loss in mango growing region of Maharashtra, Gujarat, Andhra Pradesh and Karnataka (Bhargava *et al.*, 2011). During maturation, fruit possesses non-edible sour yellow sponge like patches with or without air pocket in the mesocarp. The unusual characteristic of this condition is that at the time of picking or at the ripe stage, outward signs are not evident. These are only observable when the mature fruit is sliced. The fruit is unfitted for human consumption due to this disease. (Joshi and Roy, 1985) found that lower content of calcium and magnesium in the leaves are associated with more severity of spongy tissue. Katrodia *et al.*, (1988) has however observed the convective heat as the main cause of spongy tissue. In situ seed germination in Alphanso mango is another cause of promoting spongy tissue as reported by Ravindra and Shivashankara (2006).

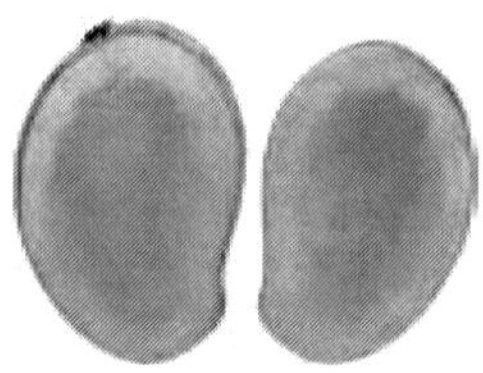

Spongy Tissue

Management

- Fruit harvesting at the 3/4th maturity stage and post-harvest exposure for 10-18 hours to low temperatures between 10-15^{0}C was helpful in reducing the condition.
- Gunjate *et al.* (1979) noted that Sod culture should be used in the orchard with Eragrostis (Spartina) cynosuroides, cover cropping, etc. to reduce direct heat radiation from the soil, which eventually minimizes the increase of ground convective heat.
- For spongy tissue susceptible orchards, the use of black poly-ethylene mulch is recommended.
- Resistant varieties such as Ratna, Arka Puneet, ArkaAruna that have the character of Alphonso should be planted in the orchard.

5. Fruit Drop

Fruit Drop

The last conservation in the mango is quite poor; despite the first large gathering of fruit (only 0.1 percent perfect flowers develop fruit to maturity). Hermaphrodite blooms in mango are too low in the quantity of functioning ovules (Davenport and Nunez-Elisea, 1997). Dahsham and Habib (1985) replied that abscission took place throughout three phases in two fruit-age months, second in a period of 60-75 days in which the fruit was pre-harvested, that is, right before the fruit was harvested. The pace of falling fruits varies across varieties. In most of the commercially cultivated cultivars, Langra is the least vulnerable to Langra cultivar. Drop of mango fruits is categorized in three groups *viz*., (i) pinhead drop, (ii) postage drop, and (iii) preharvest drop (May drop). The first two drops produce fewer financial losses for growers, but the third class is the most destructive and creates huge economic losses for farmers when their fruits fall entirely before harvesting. Because to the dissolution of embryos and ovular abnormalities, many fruits fall. Other variables including climatic circumstances, interrupted water connections, poor nutrition, competition for fruits and pests and illnesses are also significant in fruit drops.

Management

- The degree of fruit drop in mango can be substantially reduced during the entire fruit production cycle by constant and frequent irrigation.
- Orchards should be secured by planting wind breaks against desiccating winds.
- It is important to implement timely and efficient control measures against major pests and diseases.
- NAA foliar application @ 20 ppm at pea stage of fruit has been shown to be effective in controlling the drop of fruit in mango (Mishra *et al.*, 1973).
- Foliar application of GA 10 ppm, 2,4,5-TP @ 20 ppm at full bloom and panicle initiation stage respectively, whereas Ethrel @ 50 ppm and Alar 1000 ppm at flower bud differentiation was found to be effective treatment for fruit drop reduction in mango (Sharma, 2005a).

6. Clustering (Jhumka)

The development of fruit at the tip of panicles in clusters which appear as a bunchy tip known as "Jhumka" is characteristic of Mango Clustering. These fruitlets are dark green and have a deeper curvature in the region of the sinus beak. After a month of fruit setting, these fruits no longer develop over and down beyond pea or marble. Inadequate numbers of pollinators in orchards are the major culprit. Especially owing to the absence of pollination/fertilization, a number of causes are possible, such as lack of a proper pollinating population in the orchards, indiscriminate flowering by spraying of pesticides, spraying of synthetic pyrethroids, monoculture and severe weather.

Clustering

Management

- Sprinkling of pesticides should be avoided during flowering. In addition, monoculture in the orchard can be prevented by planting at least 5-6% of different cultivars in new plantings. Some branches with pollination kinds in old orchards should be top worked.
- Introduction of beehives to raise pollinators during a flowering season in orchards.
- NAA spraying @ 200-300 ppm is beneficial during October-November (Sharma, 2005a).

7. Internal Fruit Necrosis

It is characteristic in the bottom part of the browning of seed, which eventually developed into the brown, black necrotic lesion, and is characterized by the emergence of a dark green hue. The bottom half of the fruit gets necrotic in the late stage and results in the necrotic area exposing the seed longitudinally splitting the fruit. Boron deficiency is largely linked.

Fruit Necrosis

Management

- Applying boron to the soil or foliar can treat this problem. Borax @ 500 g per tree should be included in soil application at the time of fertilization in October.
- At the time of fruit set (pea size stage), foliar application of 1 percent borax is recommended, followed by two more sprays at a 10-15 days interval that will greatly mitigate the disorder.

8. Tip burn/ Leaf scorching

The characteristics of potash are comparable to those of the tips or edges of old

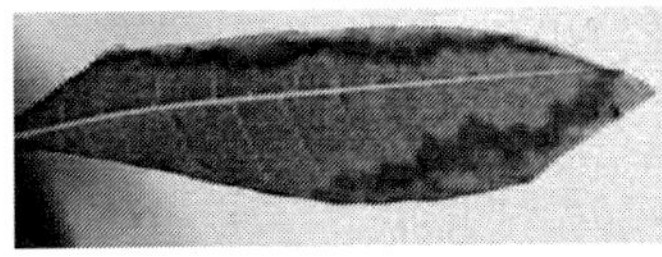

Leaf scorching

leaves. The leaves fall, thereby decreasing the tree's vigour and production. This is mainly owing to the quantity and the inaccessibility of chloride ions. This situation is more predominant in salt soils or when potash muriate is utilized as a fertilizer to meet plant potash or brackish irrigation requirements.

Management

- By gathering and burning the fallen leaves and using potassium sulphate instead of potassium muriate, it can be effectively tested.
- Acute disease can be treated on fresh flushes by 4-5 potassium sulphate foliar sprays at quarterly intervals 5 per cent (Pandey and Sharma, 1979).

9. Fruit Cracking

Cracking

In the past few years, incidences of fruit cracking have been observed and reported from different parts of the country. It is observed in mature fruits rendering them unmarketable. Fruit cracking appears to be a varietal character. Higher incidence of fruit cracking was noticed in Dashehari cultivar from the plains of North India.

Management

- Regular irrigation during the fruit maturation stage, mulching to preserve soil moisture, introducing organic manures to the soil to replenish its fertility and applying micronutrients are part of the management strategy for fruit cracking.
- In reducing fruit cracking, foliar boron sprays have been found effective. The occurrence of fruit cracking in mango can also be minimized by preventing internal necrosis and the bacterial black spot.
- Trees should also be secured from dry winds that are heavy and desiccating.

11

Physiological Disorders of Litchi

1. Poor Fruiting/ Non-Fruiting / Irregular Fruiting

Poor Fruiting

Problems are low fruiting / non fruiting / irregular fruiting in the country's litchi farming area. The potential of bear litchi has been identified as variety, although some of the bearing terminals in the present year will be more active in the next year (70-95 percent). Failure to follow proper management techniques leads to irradiation including the late administration of manures and fertilizers in regular bearing cultivars (especially nitrogen fertilizers), high frequency and heavy watering shortly before the panicle emerges and the blooming time. Cultivar China is widely known as an unauthorized carrier which may yield fruit every year when improved management methods are implemented accurately. Ireland, nutrition and environmental variables are responsible for erratic bearing of litchi. Irregular bearing or alternate bearing in combination with management techniques has also been shown to be linked with soil form. Late harvests and severe cutting and training operations might cause to non-bearing in that particular year of service.

Heavy crops often follow light crops in certain types, but output in some varieties is quite consistent each year. In certain situations, the trees begin to provide vegetative flush during the panicle and the year is decreased, while, in some cases, just half of the tree bears lengthy leafy panicle at the base which results in very few blooming ends and very small gathering of fruit. Sometimes the harvests of the mature fruits, because of heavy flowers and fruit fall at different periods, are extremely poor in spite of heavy flowers and a rich fruit set. The fruits are very rich.

Management

- Apart from other desirable characters, varieties chosen for planting must be periodic in bearing.
- The Shahi, Rose Scented, Dehradun cultivars are standard bearers, while irregular/shy bearers are China, Seedless and Late Bedana.

- Depending on the age of the trees, balanced nutrition should be properly provided to commercial orchards through the proper process of fruit harvesting and irrigation, especially at the time of the development of aril.
- Plant security steps should be followed to manage major pests and diseases.
- A good and safe flush should be plucked into the branches, at time of harvest, and immediately after harvest, during harvest, for the following year. The length of the 8-10 cm branches is necessary for excellent and safe flush.
- It is important to avoid late pruning and preparation. To give the canopy form in a semicircular way or an open umbrella shape, the tree should be conditioned.

2. Flower and Fruit Drop

Flower drop

Even after a large amount of flora and fruit harvest, inadequate yields are occasionally recorded because of a significant decline between flora and fruit maturity in lithium trees. In many different types, just a tiny fraction of it (2-18%) develops.

The deterioration in the number of flowers and fruits was believed to be due to failure in fertilization, embryo abortion, nutrition and hormonal imbalance and environmental conditions such excessive temperature, poor moisture, high winds and tears. Young trees fall below elderly ones. The overall loss of fruit among types was between 3-39.6 per cent during the first two weeks following fruiting and harvesting. Also taken into account was the specific connection between the auxin production in fruit growth and the litchi abscission.

In litchi, male functional flowers collapse after pollen production. A substantial reduction in fruit was noted at different stages because to the development of the abscission layer. During the first month following the pollination, most flowers and fruit drops develop. The fruit in the litchi is extremely high, but finally grows to a limited extent. Shortly after the fruit is set, the premature fall begins and continues until the fruits mature. In the first 2-4 weeks 70% of the fruit abscises.

Management

- By regulating winter firing, flowering cycle and flower quality, high fruit retention in litchi can be assured.
- Plant timely treatment delays blooming and raises the female flower ratio and finally the fruit set-up rate for high and steady growth.

- Honeybees are the big pollinator of litchi. The increased bee population in garden enables better pollination and fertilization, thereby improving fruit production and retention.
- At the blooming stage of the orchard, it is advisable not to make any kind of spray.
- $ZnSO_4$ spraying @ 0.2 percent stimulates healthy inflorescence, which results in more fruits 30 days prior to panic.
- 0.2% foliar boron treatment, 2-3 times during the development and growth cycle of fruit, improves fruit preservation, reduces cracking, improves fruit colour, improves taste and increases ripeness, thereby contributing to the enhanced quality output.
- Applying peanut fruit at intervals of 15 days to two leafy applications of planofix at 4ml/5 litres of water.
- Development regulator treatment such as NAA with panicle spray at a rate of 20-30 ppm, GA, 20-25 ppm and 2,4-D with 10-20 ppm is helpful in reducing the decrease in fruit.
- $ZnSO_4$ sprays on fruits that improve production will minimize fruitlet drop and on synthetic auxins, such as the 3,5,6-trichloro-2-pyridyl-Oxyacetic acid (3,5,6-TPA).

3. Fruit Cracking

It is the worst litchi disease in most of the litchi-like regions in the country or perhaps worldwide (mainly throughout Asia and the Pacific). One of the serious problems for litchi growers, mainly in early bearing cultivars, is fruit cracking (Singh, 1986). In India, it is a big problem with litchi that causes losses as high as 5-70%.

Cracking

Most of the cultivars in India are found to be prone to fruit cracking. As a result of development, cell division and fruit scars are inconclusive and sometimes disintegrate when the aryl grows quickly before harvesting. After irrigation or heavy rainfall or relative humidity rise, this can occur. After a long distance the temperature and humidity in the atmosphere, cracking becomes more apparent, simultaneously heavy rain orchards or sudden heavy rains. Nutritional deficiency affects fruit production, resulting in fruit cracking. Boron-deficient soils may also have calcium activity that may be associated with fruit cracking. Insect or mechanical skin damage during the development process of pericarp can also cause fruit cracking. An association between fruit cracking and endogenous growth substances was observed in high abscisic acid (ABA) concentrations and low gibberellin levels in the cracked fruit pericarp, seed and aryl. Cell expansion, insects, hail, and sun damage the skin and crack the harvest. With the rest of the fruit, these weakened areas cannot expand, causing the skin to break apart. Lychee cracking, which coincides with the highest stage of aryl development, occurs mainly after a break in the color of the fruit. Thick-skinned cultures are less sensitive than Thin-Skin cultivars, and those with inadequate management techniques are more likely to develop this early maturing disease.

Factors Influencing the Fruit Cracking

1. Ecological Factors

- ***Temperature influence:*** In the fruit cracking ratio, temperature has a very major impact. Higher temperatures and very low humidity or precipitation in dry and semi-arid regions encourage cracking. The cracking generally increased linearly when the temperature increased. Other factors such as cell wall permeability and metabolic activities can also be affected by temperature.
- ***Wind***: The presence of loo/heat winds during the summer months is a typical phenomenon in Northern India. As such heated wind from the litchi fruit surface flows, because of water loss, it becomes difficult and inelastic and leads to cracking in litchi fruit in case of sudden fluctuation in moisture level.

2. Features of Fruit

- ***Maturity:*** Over ripeness, which contributes to fruit epidermis cracking?
- ***Fruit size:*** Large fruits are usually thought to be more vulnerable to cracking than smaller ones.
- ***Fruit firmness:*** Corporate meat fruit, such as litchi, has shown more sensitive than soft fleshly fruit to breaking of the fruit. Excessive water absorption leading to skin bursts causes fruit cracking.

3. Lack of Management of the Orchard

- ***Moisture stress:*** After a lengthy dry heat, humidity and heavy rainfall or irrigation, the water supply to plants may suddenly and highly fluctuate, causing the fruit to break down.
- ***Nutrient:*** Cracking is a matter of boron and calcium deficiencies in pomegranate and litchi.
- ***Insect-pest and disease***: The sunburns, which ultimately acquire the shape of elevated spots, are a small, black water-forming area in litchi. These zones cause longitudinal breaks on the fruit and start to splash out.

- ***Bagging***: The water is also a restorative operation to escape the sunshine from the plant surface as it is loosened by transpiration by stomata
- ***Early picking***: Early fruit harvesting is also a corrective method to alleviate splitting. This does not allow excessive or excessive ripening, causing breaking of fruits.

Management

- The best possible approach to mitigate rainfall in fruit orchards is to pick the site. The right site should be with little or no rainfall throughout the close harvest period. Choose crops with resilience to rain-related fruit cracks. The physical growth of fruit from heavy precipitation is covered by water resistance.
- The use of liquid calcium formulations @ 2 m/l and Gibberellins @ 20 ppm lowers the activity of cellulose and hence minimizes cracking.
- Sprinklers of 2,4-D and NAA reduce cracking at 20 ppm or 20 mg/litre.
- Constant moisture and appropriate moisture is necessary during fruit maturity. Irrigation at 30-40% of available soil humidity is extremely useful to minimize the cracking of the fruit. Under the canopy area, dropping and micro sprinkler facilities have been reported to be beneficial in preventing fruit cracking.
- Mulching can play a significant role in regulating the temperature and humidity of the root zone. In these frontier cultures, planting maize or sugarcane and frequent watering around litchi orchard generate a beautiful microclimate that prevents fruit cracking.
- The plantation of wind breaks around the orchard gives protection from hot winds. There should be sufficient wind breaks along the boundary and the wind route should be at the appropriate angle.
- At the beginning of aril growing with adequate soil moisture in the root area, boron spraying in the form of borax, or boric acid @ 2g/l considerably prevents fruit cracking. The needed wind breaks are a series of big trees, including seedling mango and jamun.
- Cracking resistant cultivars must also be established or discovered as just one cultivar, i.e. It is stated that Swarn Roopa, produced in Ranchi, Jharkhand (India), is tolerant of this disorder.

4. Sunburn

Sunburn, sometimes called lesion browning or pericarp necrosis, is a major issue in litchi-producing areas. The occurrence and gravity of sunburn disorder are determined in some growing locations by climate conditions and cultivars. Sun-blasting causes damage of up to 0.9-19.13 percent across diverse variants.

Sunburn

In addition to environmental variables, this condition also includes varietal, hormonal, nutritional and soil humidity components. The physiological state of PPO (poly-phenol oxidize) is also associated with cultivars, Sunburn is marked in unsafe olive groves, receiving/exposed to sandy/sandy loam soil, or light soils, with a high temperature (> 40°C) and very little RH (<50 per cent). The damage is direct and sunburnt/sun scalded in the case of higher temperature and the contact/exposure of the tissue. More is also seen in early ripening cultivars in the Sunburn crisis. Fruits on shaded trees are less affected than those that are more open to sunlight. Higher calcium translocation in the area of the pericarp has also been found to encourage sunburn disorder.

The area of the fruit skin faced by direct sun rays is littered with brown blotches in the case of sunburn. More than half of the surface area is discoloured, and light brown is blotchy in extreme instances. In a few of days, the blockages get serious, and the tangy area dries to hinder the aril growth and ultimately ruin the market worth of the fruit. Symptoms happen on the southwest more than on the north-east, since the fruit on the latter side remains virtually in shadow, except during the early hours of the day. Due to the position of the fruit on the trees, fruit producing might suffer from this disease under comparable climate circumstances.

Management

- Irrigation minimizes sunburn at regular intervals throughout fruit development and maturity.
- Planting of windbreaks around the orchard covers the desiccating warm wind, therefore ensuring that sunburn is low.
- Sprinkler irrigation provides moisture during hot hours, cools the orchard's environment, and therefore minimizes sunburn.
- The only advantage of light and sandy soil is light water with increasing frequency (4-5 day interval).
- Maize or sugar cane planting and daily watering in these border crops provides a pleasant microclimate to minimize sunburn.

5. Retarded/ Underdeveloped Fruits

Not well-known yet with low grade fruits or badly prepared fruit it causes significant losses. The fruit is smaller than normal and also has extremely little amount of juice and aril. This sort of fruit has been shined with a faint pale green

Underdeveloped Fruits

and only then emerges from the question of prematurely fallen fruits in a dull reddish hue. Seeds originally develop at high rates while litchi is being grown, followed by membrane mesocarp and aril which expand rapidly in the final stages of fruit formation. Often, during these phases, no exposed part of the testus leads to the formation of delayed or undeveloped fruits aril is incomplete. Failure to adequately fertilize owing to fewer visits by the pollinator might result in improper fruit preservation and delayed fruit production. Insect infestations and physical jerk often produce delayed or undeveloped fruits. The lack of moisture available during the fruit production is one of the reasons of retarded/underdeveloped fruits. It is more natural in poorly preserved orchards with limited pollination.

Management

- Proper nutritional control of the orchard must be enforced.
- The amount of fruit bee colonies placed during blooming enhances the fruit set with greater fertilization leading to normal and healthy fruit growth.
- At the early morning hours of the day, the spraying of flat water was very beneficial in improving fruit development and quality at the advanced phase of growth and development.
- Irrigation device sprinkler arrangements have also been shown to be effective in minimising this condition.

6. Black Spot

Black Spot

The black spot in litchi is not common; however, the quality of the litchi harvest has been proven to cause infestation and harm. The illness was mostly detected in the proximity of urban regions in established plantations under Indian circumstances and no more account is made of the disease occurring in litchi. It is produced by smoke gases including sulphur dioxide, acetylene and carbon dioxide which are harmful in the case of mango. The black mark is shown when the fruit reaches a particular size and is retained during the development phase, from late March to mid-May. The first symptom arises at the distal end of the resulting fruit of a tiny etiolated area, which is practically black and covers the distal end with its typical green colour of the pericarp fruit. The infected fruits display discoloration and sluggish growth overall, which lead to very poor growth of aril, seed, size and colour.

Management

- The condition can be minimised by using adequate nutrition and watering, particularly during the growth of fruit and aril.
- A month before flower panickle start or boron (borax @ 0.2 percent) throughout the fruit growth phase, the issue can be mitigated by spraying with micronutrient such as zinc.

7. Pericarp Browning

Pericarp browning is the first visible sign of deterioration. The first few days following harvest of pericarp dehydration usually leads to browning. This problem has been observed in various locations of litchi production including the development of necrotic patches that can extend to a large proportion of the fruit. Once a little bit of the fresh weight of the crop is gone, fruit starts to discolour. The pericarp is entirely orange, under the initial fresh weight of 50 percent. Pericarp browning takes place at ambient temperatures between 20 and 30 °C within 24 hours. The lack of litchi water contributes to the red shell's brown patches (pericarp). Other circumstances cause a fruit to brown, including various types of mechanical stress, microbiological and assaults by insects, and high temperatures (predicating at harvest, sliding off the fruit in a raw picking bag, drop fruit from short heights). In short, fruit browning is likely to hasten the breakup of the cells. This happens owing to injury to the thin mesocarpal parenchymatic cells (due to unknown causes). A hypersensitive reaction characterized hassles this injury to the epicarp and endocarp, resulting in the necrosis of these cells. The increased activity of the enzymes of polyphenol oxydase and perioxydase and oxidation of ascorbic Acid is related with skin-browning. Even if aril is not damaged, it can cause considerable losses; the fruit cannot be marketed. In severe situations, the pericarp may break. With the application of fungicides and cooling, Litchi fruits have a shelf life of around 30 days. However, their shelf-life is quite short when taken from storage because of the pericarp browning and rotting of the fruits.

Management

- The packaging into (plastic) moisture-proof bags and pinpoints reduces water loss and slows the browning rate considerably. The less security, for instance papier, wicker baskets and carton, creates more permeable obstacles.
- Furthermore, low storage temperature retards browning. Low temperatures may be sluggish to evaporate slowly and to breathe slowly.
- A balanced environment of 3 to 5% O_2 and 3 to 5% CO_2 also demonstrated sluggish water loss.

- If correctly applied, sulphur is also anti-pilzing. Sulfur dioxide fumigation lowers pericarp browning substantially, although some nations are not allowed to do this because of worries about sulphur residues in sprayed fruit. Pre-storage pathogen control is a way of avoiding browning pericarp and fruit rotting alternatives to SO_2 fumigation.

12

Physiological Disorders of Apple

1. Bitter Pit

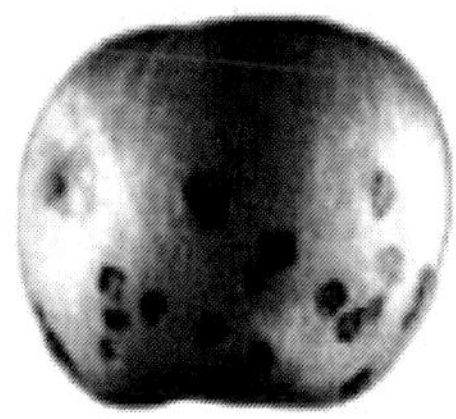

Bitter pit

In the flesh of the fruit, tiny, 2-10 mm (depending on the cultivar) brown blemishes form. Below the surface, the tissue gets black and corky. At harvest or after cold storage, the skin develops depressed patches on the surface. These tend to develop as water-soaked patches on the skin around the calyx. Overall, these patches became darker and sunken than the skin around and developed entirely after one to two months in storage.

Fruit on strong, green, upright branches is more likely to grow bitter than fruit, which grows in spurs or on horizontal wood near a tree's main structure. Young trees that merely come into existence are the most susceptible. Immature fruits are more susceptible than fruits which are harvested when the harvest is ripe in the bitter pit. Older trees, less strong and more crop-load, decrease their sensitivity to bitter pits with the exception of delicate crops.

Highly Susceptible Cultivars

Gravenstein, Red Delicious (California), Baldwin, Boskoop, Bromley's Seedling, Cleopatra, Cox's Orange Pippin, Grimes Golden, Merton, Rhode Island Greening, Worcester, Stayman (Canada), Sturmer, White Winter Pearmaine, Prima Starking, Starkrimson, Marigold, Northern Spy, and York Imperial. Granny Smith, Yellow Newtown, Jonathan, Golden Delicious (California),

Less Susceptible Cultivars

Rome Beauty, Golden Delicious (Canada), Haralson, Lobo, Winesap (California), Macintosh, Spartan, Red Gravenstein, Stonetosh, Gala and Fuji.

Management

- Before collection the presence of bitter pits before storage is monitored by calcium sprays and calcium dips. 45 days before harvest, followed by repeated spraying for 15 days, the plants should be sprinkled. The post-harvest dip should be given for 1-2 minutes before storing.

- Stop the sprays of nitrogen and magnesium in summer, fertilizer treatments which reduce soil pH or produce over vigour and volatility of soil moisture.
- The bitter pit will increase early thinning and over-thinning.

2. Brown Heart

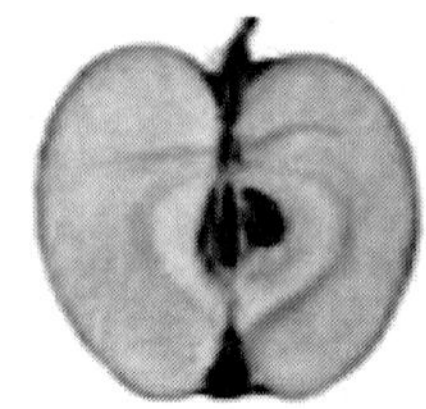
Brown heart

Large and over-ripe fruits are associated with this physiological condition. As the CO_2 concentration in storage rises above 1 percent, it may also occur. The signs manifest in the flesh, usually appearing in or near the heart, as brown discoloration. Brown patches have well-defined edges and may contain dry voids developed due to drying. Symptoms range from a little area with a brown body to a complete body with a white healthy border underneath the surface. Symptoms begin early in storage and can increase their severity with longer storage duration.

Management

- Harvesting of over-ripe fruits should be prevented. In the case of Controlled Atmosphere (CA) storage, the fruits should be harvested at optimum ripeness.
- To decrease the production of brown heart incidence, the CO_2 concentrations in CA should be below 1 percent.

3. Cork Spot

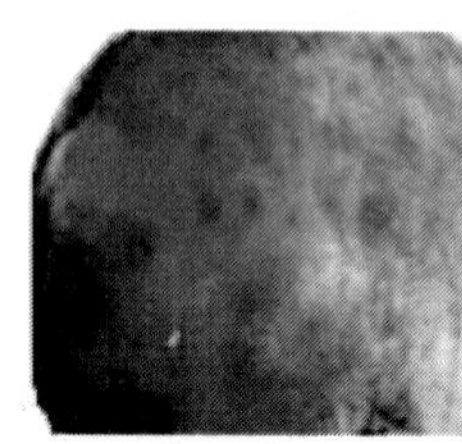
Cork Spot

As a small blushed region on the surface of the fruit above the infected brown patch, the initial signs of this physiological condition appear. The affected tissue is typically much tougher than the healthy tissue. Boron and calcium deficiencies are often found to be responsible for cork spot growth.

Management

- Proper nutrient management helps in preventing this disease, especially boron and calcium.

4. Scald

Scald

This physiological state is a serious concern for apple farmers. Apples, civilizations, and the environment vary in their sensitivity to this disorder. The occurrence and degree of scalding is promoted by hot, dry fruit before to harvest, immature fruit during harvest, high levels of nitrogen and low calcium in the fruit. Often, insufficient ventilation in storage rooms or packaging boxes makes these conditions easier. Due to warmness, uneven brown patches of dead skin form within three to seven days after cold storages have been removed. Warm temperatures do not scald but allow for symptoms from past injuries during cold storage. Signs might be seen if the damage is serious in cold storage.

Management

- Harvesting at sufficient maturity and ventilation in cold storage helps reduce the occurrence of scalding.
- Applying an antioxidant immediately after harvesting is the most common process used to control scalding. It is normal to use diphenylamine (DPA).
- Antioxidants should be used for optimal regulation within one week of harvest.

5. Water Core

This physiological state is encouraged by large fruit leaf-to-fruit ratios, high nitrogen-based fruit boron concentrations, low calcium fruit levels, excessive dilution and exposure to high fruit temperatures. This form of condition is mostly prone to large-sized fruits. Disorder results in the creation of water-soaked regions in the flesh at the pre-harvest stage.

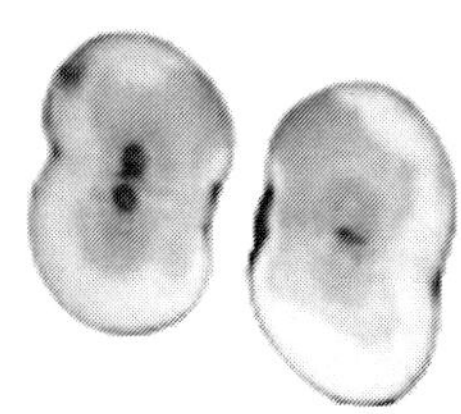

Water Core

These areas are robust, glassy in appearance and may only be seen externally if the illness is very serious. Badly infected fruit might taste, and smell soured. Water-soaked patches might be found close to the heart or on the whole apple. If the symptoms are mild to moderate, they may subside fully with storage.

Management

- Avoiding delayed harvests is the most efficient way to lower the incidence.
- As fruits reach the stage of maturity, fruit samples for water core growth should be studied. Fruit should be harvested before comprehensive production of the water heart.
- No lots of moderate to serious symptoms of water core should be maintained in the controlled storage environment (CA), but should be marketed promptly.

6. Sun Burn

The sun's intense heat causes this physiological state. In general, fruits are damaged on the southwestern side of the tree. Water stress can also enhance the likelihood of sunburn. The first indications of this are white, tan or yellow patches on the fruit exposed to the sun. The damaged fruit regions are dark brown with serious skin damage before harvest. Sunken and spongy might be these places. The sun-exposed fruit develops severe sunburn after harvest.

Sun burn

Management

- The best control strategy is to prevent sudden exposure to excessive heat and solar radiation from the fruit.
- Proper training of the tree and pruning are important. To prevent unnecessary sunburn, summer pruning must be carefully undertaken.
- To minimize heat stress, pruned orchards should be irrigated periodically.

7. Russeting

A major concern of fruit growers is the ruse of apples in a humid climate. Russeting happens soon after the dropping of the petal. The cultivars of apples with thin cuticles are more vulnerable to russeting. On exposed fruits, it is typically found than on fruits remaining in shade. Russeting can also be caused by frost during flowering or at the early stage of fruit development. Russeting leads to fruit skin rupture and crack growth.

Russeting

Management

- Choosing less vulnerable clones and appropriate irrigation, manuring and effective management of pests will minimize russeting.

8. Fruit Drop

Most commercial apple varieties exhibit 3 cycles of drop in fruit, namely early drop, drop in June and drop in pre-harvest. The early drop is deemed natural and is due to a lack of pollination and competition for fruit. The June drop is caused by moisture stress and environmental factors. These two declines do not cause significant economic losses or are effectively regulated by artificial means. As the fully grown marketable fruits abscise before harvesting due to the decrease in auxin levels, the preharvest drop causes significant economic loss.

Fruit drop

Management

- The pre-harvest drop can be managed by NAA (15 ppm) spray 20 days prior to the predicted drop in fruit or 20-25 days prior to harvest.

13

Physiological Disorders of Grapes

1. Hen and chicken

Hen and chicken

In grape growing areas in many parts of the world this illness is a very popular disorder. In this disorder, a bold beer is surrounded by shoot berries, and the name of the hen and of the chicken disorder is like a hen, as bold berry and chicken. The fruit has a sour taste. The signs are the death of rising tips, falling leaves and the fragility of new shoots. With the development of interveinal chlorosis between the edges, the leaves may be deformed, and this is particularly evident after fruiting. It mainly happens to the deficiencies of zinc and boron micronutrients and improper fertilization. So, foliar application of zinc and boron lead to controlling this malady.

Management

- Sprinkling 0.2 percent boric acid per week before the blossoming, while another with full bloom controls the problem well.

2. Blossom-end rot

At the blossom-end of the berry, a black sunken spot grows which later spreads with the area around it water-soaked. Defective diet and assimilation of calcium tend to be the cause.

Management

- Percent calcium nitrate spray will repair it.

3. Uneven ripening

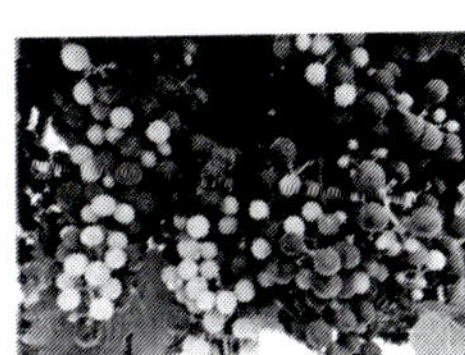

Uneven ripening

The emergence of green berries is regarded to be incongruous in a full cluster of coloured grapes. It is a varietal trait and problem with Bangalore Blue, Bangalore Purple, Seedless beauty and Gulabi grapes. This problem varies from one bunch to another in a set. This is typically explained by inadequate leaf area and the non-reserves available for a growing group.

Management

- Inconsistent maturation, such cluster thinning, girdling and growth regulators can be minimized by cultural practices.
- In order to reduce the issue, Ethephon treatment (250ppm) is advised during colour breaks.

4. Water Berries

Water Berries

The afflicted berries are watery, mushy and flabby during maturation because of the disturbed flow of sugar and other ingredients into the berries. The berries impacted tend to seem more metal, opaque or dark green; the berry colours may be lighter and more varied in both the red and black types. This occurs from the collapse of the cluster structure and necrosis of the sustaining cluster parts. The passage of water may eventually be blocked, and the fruit may become shriveled and dry. The first symptom is a small (1-2 mm) black patches of the cap stems (pedicles) and/or other sections of the cluster system. Such patches are necrotic, slightly sluggish and grow in additional locations. The cluster's tops, shoulders, and top sides are primarily impacted; the entire system may be implicated in significantly impaired clusters. The necrotic cluster stem may have a dark brown to purplish-black or black colour.

Management

- To minimize water berry formation, excessive irrigation and nitrogenous fertilizers should be avoided during berry growth.
- In water berry prone vineyards, foliar nutrient sprays of nitrogen should be avoided. Similarly, near the beginning of or during fruit ripening, ammonium or urea fertilizers should not be applied through drip irrigation.
- Trellising and canopy management practices can be helpful, such as shoot thinning that minimizes shading within the canopy.

5. Cluster Tip Wilting

Cluster tip wilting

The conductivity of the rachis is compromised by light brown lesions on the apical tip of the rachis. This results in the rachis at the tip of the bunch shriveling and drying. The tip of the bunch dries up to 30-40 percent in extreme cases, leaving rough, tiny and light brown berries at the tip. To this condition, Thompson Seedless is more prone.

Management

- To minimize excessive crop load on the vines, cluster pinching or berry thinning is recommended.
- Ensuring sufficient irrigation during the growth of berry bunches and shielding them from direct sunlight.

6. Shot Berries

They are produced due to the delay of a few flowers in pollination and fertilization or due to insufficient carbohydrate flow into the set berries. The known causes for shot-berry formation are Boron deficiency, incorrect stages of GA application and girdling.

Shot berries

Management

- Similarly, the use of GA should be ensured at the right stage.
- Boron or Zinc deficiencies should be corrected.

7. Pink Berry

Some berries in the bunch spontaneously grow a pink colour as the bunch reaches maturity. The pink colour switches to a bland red colour that makes the bunch unattractive.

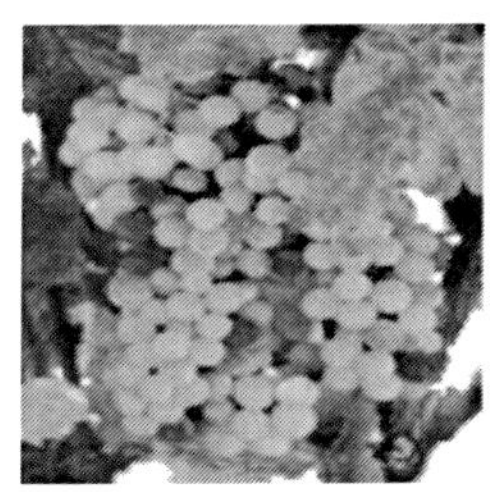

Pink berry

Management

- This condition may also be caused by indiscriminate use of Ethereal for berry colouring.

8. Bud and Flower Drop

A number of variables are responsible for the illness, such as inadequate nitrogen use, inaccurate fertilization, ambient temperature, high crop loads, uneven ripening and deficit in endogenous auxins at a particular berry production stage.

Bud and flower drop

Management

- Make a 0.5 cm wide trunk bird approximately 10 days before full bloom and make a better gathering of berries.
- At the Veraison point, 500 ppm Ethrel should be added.

- 10 days before ripening, dipping bunches in NAA 100 ppm decreases berry drop.
- Strong irrigation during flowering should be avoided.
- GA, 20 and 50 ppm and Parachlorophenoxy acetic acid 20 ppm applied at 18 days before anthesis reduce the flower bud drop significantly in 'Thompson seedless' grape (Jawanda *et al.*, 1974).

9. Poor Cane Maturity

Shoots fail to develop in this form of disorder, and their barks remain green until late in autumn. These shoots turn pink-red in winter due to low temperatures. In vineyards, where shoot growth is vigorous and thick, it is more serious; vines are closely planted and excess nitrogen and irrigation are given.

Poor cane maturity

Management

- Judicious pinching of shoots to control excessive vegetative growth.
- Some of the suggested remedial steps include shoot thinning 30 days after summer pruning to avoid mutual shading of the shoots and encourage light interception.
- In the 40-70 days following back pruning, avoiding unnecessary irrigation and nitrogenous fertilizers helps to overcome cane immaturity.

14

Physiological Disorders of Bananas

1. Choke throat

Choke throat

It's because it impacts the active development of the plant at a low temperature. Yellowing of the leaves and severe destruction of the tissue. In cases of flowering plants, the stalk carrying the bunches freely lengthens and drapes down the complete inflorescence. Bunch is normal, however when the flowering time synchronizes with cold temperatures, the bunch cannot come up appropriately from the pseudostem. The distal section of the inflorescence is drawn and the basal section in the mouth is trapped. It is hence known as Choke's throat. In fact, the maturity of the bunch is delayed by 5-6 months rather than 3.5-4 months for harvest.

Management

- The adoption of low-temperature types and the use of eucalyptus as a shelter belt to control the effects of cold wind needs chocolate management.

2. Chilling injury

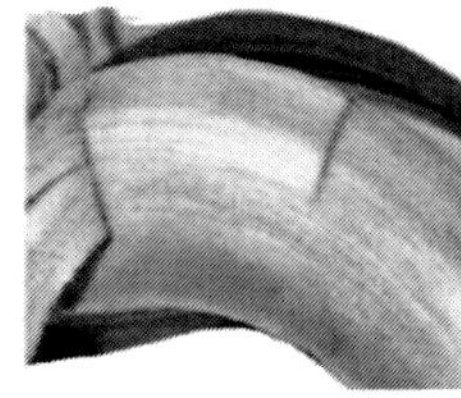

Chilling injury

Banana chilling happens if the pre- or post-harvest temperature is below 14°C. The symptoms include uneven ripening, watery dark skin patches and dull yellow-smoky yellow ripening of the fingers. Brown strips are also observed in the vascular bundle of the sub-epidermal layer. These are the consequences of enzymatic oxidation by dihydroxy phenylalanine.

Cold injury losses may be greater than accepted after harvest. Very often, when the product is still in the refrigerator, the symptoms of a cooling injury may not be obvious; symptoms appear later, only after the product has moved to a market where the temperature is higher. Symptoms of a cold injury become apparent at the harmful cooling temperature or after transferring the fruit to a non-cooling temperature after several days of storage. Green fruit of the banana develops surface lesion, such as the first epicarp boiling, and, if the refreshment is not excessive, the colour of the skin is dull yellow to gray-yellow or grey, which usually applies under the conditions of the export of bananas.

Management

- Stop storing fruits below 13°C. (55°F) this disease may readily be overcome by Musa cultivars stowed at a temperature above the threshold level.

3. Kottaivazhai

Kottaivazhai

It is a major disease of the Poovan banana type, which decreases yield by 10-25%. The indications include notably conical and unfilled fruit, many of which have undeveloped seed systems that have a conspicuous central core that does not eat the fruit. On the surface there are striking, striation and blocking in the pseudostem. The bunches are positioned at an angle above the horizontal position. The pericarp is smaller, the chamber is bigger than usual, yet pollen is sterile, shrinky, twisted and fragmented.

Management

- 20 ppm (1g in 50L of water) 2, 4-D can be sprayed. It can be handled. It is possible to fully regulate and make the fruit good in quality by adding this seediness of the fruits.

4. Finger drop

Finger drops

The pedicle is broken and fluffy in a state. The fruit may be readily separated from the bunch throughout the ripening period. The finger falls into "physiological weakening of the pedicel which separates the fruit of one hand from the crown" extremely readily. This condition is related with a decrease in the pedicel without the existence of a preferred abscission zone which shows a weakening of the skin. This disintegration occurs in the pedicel (linear tissue), pulp (not-linear tissue) area of the joint. A finger drop assessment might be done by submitting a matured cluster to a 3-5 manual shaking. By tracking the number of fingers dislodged, the proportion of the fingers that fall per cluster may be calculated. The strength (N) needed to separate the pedicel of the finger is another method.

Finger drops occur during banana maturation. Variation-dependent susceptibility was reported in the diploid, triploid and tetraploid variants. Tetraploid cultivars appear to have been more susceptible than others. The severity of this physiological condition is affected by external variables after harvest. High relative humidity stimulates finger drops. However, this process does not seem to be influenced by the water content in the peel. High temperatures and ethylene create a bad neck. The physiological age of the hands has an influence on sensitivity: it is more advanced. The bad neck might have three probable reasons: biology, nutrition and circumstances after harvest.

Management

- Provide the fruits for maturation with ventilation and a temperature of 18°C.
- Use ethylene (1ml/L) at 14-18⁰°C and 90-95% relative humidity for a 24–48-hour period for fruit ripening.
- Using cardboard for wrapping the fruit bunches.

5. Peel Browning

The browning of peels is an individual phenomenon that occurs throughout the maturity stage when the fruit is processed. Some kinds seem to be more sensitive than others. The brewing of peel is due to low temperature or relative moisture voltage. Peel brewing may be evaluated by storing mashed bananas in the room at ambient temperature (20°C) and low relative humidity (50 percent). The treatment of fruit may induce browning if the variety is susceptible. The browning of the banana peel is reflected in the browning of the peel resulting from manipulation of the skin. When bananas are treated, the degree of browning rises every day under dry conditions. Symptoms grow over the entire surface of the banana.

Management

- The stored fruit will be delayed for the browning of the skin with high relative humidity (95%). Post-harvest browning appears to have been connected with the drying of the peel. The air extraction method and the use of macro perforated polyethylene bags can regulate relative humidity of the atmosphere during shipment and maturation. in practice.

6. Peel Bruising

During the marketing of bananas, the fruit is handled on the market stands, and there are mostly damage to the bananas. Mechanical contact between the fruits and other corpses has a negative impact on contusions. A tiny stainless steel ball (19 mm) is used in the peel bruising of a guiding tube, which is collected at an optimal thermal value from a range of standard height at various energies.

The tendency to bleeding is not a physiological condition, because mechanical damage can lead to a combination of an impact, and especially a breakdown of the skin. Bruising helps to create a brown black look on the fruit surface. This happens fast after the injury and extends over the whole field of impact. Symptoms are noticed on the banana skin, however there is no effect on the fruit content.

Management

- Adequate product conditioning may prevent blurring of peel. As greater turgidity allows increasing the threshold at which the banana skin is vulnerable to bruising, fruit in the morning should be harvested and the next actions taken to limit loss of water are to be strengthened.
- Reducing the temperature of the after harvest chain might decrease blurry damage.

7. Peel Cracking

Cracking

The vulnerability of peel cracks are basically affected by genetic, environmental, morphological and physiological variables, however after harvesting, peel splitting is largely due to environmental stocking circumstances. It only happens between the third- and sixth-day following maturation induction. High relative moisture usually promotes cracking of peels. Some cultivars are more sensitive than others, such as 'Mas (AA).' When kept in a perforated polyethylene film in a carton, green bananas collected with the optimal thermal quantity may be measured by a scattering of the peel. Ethylene exposure induces maturation. Polyethylene film is then closed to divide by saturated humidity conditions (90-95 percent RH). In comparison to the overall quantity of bananas, the division proportion was determined as the number of split bananas. Peeling is a longitudinal rupture of the peel, showing the pulp. Peeling is a state of physiology. It usually originates around the pedicel and, more particularly, at peel intersections between two flat segments.

Management

- The application of dry moisture conditions, such as control of the amount of perforation of bags of polyethylene or of moisture in the matured or storage rooms during ripening can be used to control peel cracking
- Pre-harvest conditions also appear to affect the appearance of the peel breakage, such as calcium deficiency
- Provide ventilating fruits with a temperature of 18°C for ripening.
- For fruit ripening at 14-18°C and 90-95% relative humidity for 24-48 hours, use ethylene (1ml/L).
- Use cartons to wrap the bunches of fruits.

15

Physiological Disorders of Citrus

1. Fruit Splitting

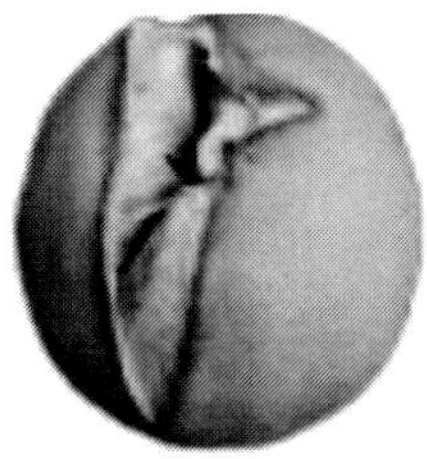

Fruit Splitting

This is primarily a physiological condition and, after heavy rains or heavy irrigation during hot weather, is mostly due to high atmospheric humidity. Two forms of splitting have been found, namely radial and transverse. There is more frequent radial cracking than transverse cracking. While splitting down to the inner core is very unusual, partial splitting is more prevalent. Organisms like *Aspergillus, Aalternaria*, *Fusarium*, and *Penicillum* frequently infect the fractured surface of the fruit, which induce partial rotation and early fruit drop.

Management

- The illness is mitigated by frequent and mild irritations during dry and warm times and by early fruit plucking shortly after its maturation.

2. Fruit Drop

Fruit drop

The most fruiting stage is(i) immediately after the fruit is set at the point where it can be grown, which is one month after full growth, which refers to the fall after the second set (ii) of Heavy fruit drops occur at the beginning of the hot summer months during May to June, referring to the fall of June; and (iii) the fall before the harvest or the fall before the harvest in the ripening season. Period between August and January high summer temperatures, excess or insufficient moisture in the soil, lack of nutrients such as zinc and fungal diseases such as late and late red colour and attacking them.

Management

- NAA (10 ppm) is sprayed monthly intervals from August to October, in order to decrease the pre-harvest decline.
- Application of 2,4-D 10 ppm in conjunction with ureofungin 20 ppm in the first week of September provides effective control of physiological and pathological decreases of pre-harvest citrus fruits.

3. Oleocellosis

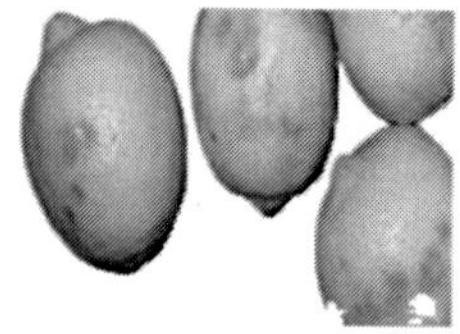

Oleocellosis

When oil cells get ruptured during harvesting or during handling from the field to the pack house, it is caused by rind oil release. In 'Mrig' bahar, it is more popular in crops taken and harvested during February- March. The occurrence of oleocellosis is minimized by diligent harvesting and handling. The skin is discolored with rind oil from ruptured cells, making the fruit unmarketable.

Management

- To reduce its occurrence, the fruit must be treated at the temperature between 18 and 20°C for 12-24 hours overnight, before the fruit is moved from the field into the pack house.

4. Granulation

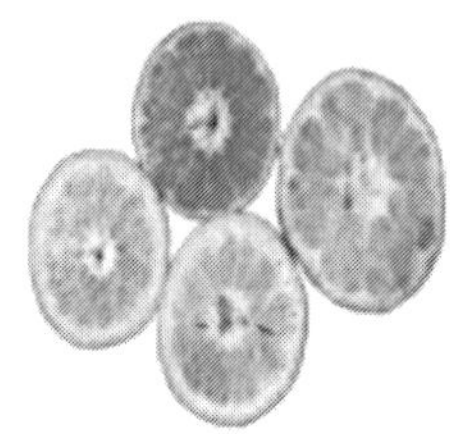

Granulation

Granulation, especially under North Indian circumstances, is a serious citrus problem. At the stem end of the fruit, this abnormality damaged juice bags that become tough and dry, the fruit grey, growing in size, taste flat and insipid and grainy in nature. In granulated fruits, there is less extractable liquid that turns into gelatinous weight. This helps to produce more rags and therefore a low pulp/pulp ratio (Zong *et al.*, 1979). The words granulation, crystallization and dry end are employed to explain this intricacy. It is considerably more frequent in larger fruits than in tiny fruits, smaller than old trees and wet than in dry regions. Almost all citrus fruits are granulated, although sweet oranges and mandarins are more severely impacted than others. (Singh, 2001; Sharma and Saxena, 2004). The illness is linked to various elements such as root and variety luxury growth, regular watering, plant tissue mineral components, harvest times, sunshine exposure, and so on. Areas with a high occurrence of granulation produce high levels of Ca and Mn in plant tissues and low levels of P and B.

Management

- The frequency of granulation might be cut back to 50% by administering 2 to 3 NAA sprays (300 ppm) in August, September and October.
- In October and November, spraying of GA 15 ppm accompanied by NAA 300 ppm also decreases granulation.

5. Citrus decline

Citrus decline

Citrus decline is not a single condition but is an integrated manifestation of a plant's multiple disorders. The term decline actually signifies continuous dying of the twigs. Tree grows well for the first 5 to 6 years, but thereafter, they show retarded growth and become unproductive. In severe cases, they may wilt and die. The improper and inadequate nutrition as well as low organic matter, excess of Fe, higher uptake of Mn and non- availability of micronutrients like Zn, Cu, Mg and boron aggravate tree decline. Apart from these the excessive irrigation, water stagnation and use of saline water also leads to citrus decline.

Management

- Due to the prevalence of soil salinity, viruses and nematode problems in different regions, it is highly desirable that rootstocks be thoroughly evaluated before use.
- Proper drainage, avoiding excessive irrigation and following clean cultivation is important to reduce the incidence of this disorder.
- Proper manuring and use of resistant rootstocks and disease-free certified bud wood maintains the growth and vigour of the plants.

6. Puffing

Puffing

Puffing disorder of citrus fruits occurs before it reached to maturity. The causes of this condition have been connected to the control of water exchange by peel. High relative humidity therefore increases the appearance and intensity of puffing, particularly in the time of drought, along with elevated temperatures at the fruit colour break point.

Management

- Proper drainage, avoiding excessive irrigation and following clean cultivation is important to reduce the incidence of this disorder.

16

Physiological Disorders of Guava

1. Bronzing

Bronzing is a complex nutritional condition in guava (*Psidium guajava* L.). It manifests in the interveinal tissue of older leaves as bronze or copper colour development, while terminal leaves remain green. This is attributable to the inadequacy of elements P, K and Zn. Bad management and low soil fertility coupled with soil acidity are often due to it. In extreme cases, complete defoliation, brown-coloured patterns on the fruit skin and decreased yield are noted. Bronzing in guava was more severe on Entisols, followed by Vertisol and Inceptisols.

Management

- Avoiding guava plantation in acidic soils or soils with high water table along with good cultural practices including balanced irrigation.
- Foliar 0.5% diammonium and zinc sulphate applications reduced bronzing of guava over two months, in combination, weekly intervals.
- Sprays with pre-flowering enhanced fruit output and fruit size by 0.4% boric acid and 0.3% zinc sulphate.

2. Fruit drop

The decline in fruit in guava is a serious concern as it results in a loss of fruit of about 45-65 percent due to environmental factors. This occurs due to poor fertility status of soil, poor moisture content in the rhizosphere during fruit growth, hormonal imbalance and attack of insect pest and diseases.

Fruit drop

Management

- Proper plant safety and GA_3 spraying methods have been found to be effective in minimizing the decline in fruit in guava.

3. Die back

The young and elderly trees are harmed; some arbours are usually susceptible to assault, withering from top to bottom of the lower branched, dark browning and necrotic propagation of the rising end. The typical lesions grow and migrate towards healthy regions; exfoliation of infected branches makes them appear unfertile in the joining of sick and healthy areas.

Management

- Preservation of soil pH reduces the death rate of trees by applying 1.8 kg of lime or gypsum.

17

Physiological Disorders of Papaya

1. Skin Freckles

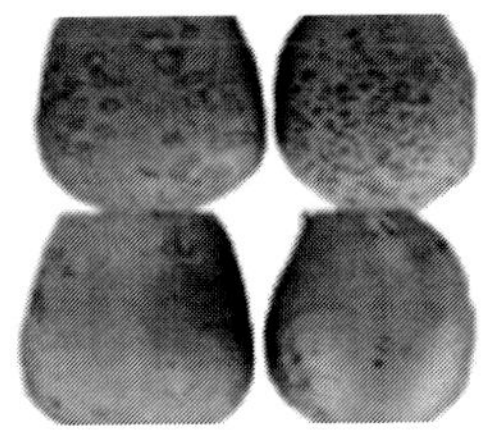

Skin freckles

Twitching is a condition occurring predominantly on the surface of the fruit which is exposed to direct sunlight; however, it is not well understood what the source of the disease is. Growth in immature fruits cannot be detected during later stages of fruit maturity. On cv. Golden Solo Party there are a few examples. This is termed the "frog skin" or "fox skin." It shows on the fruit's surface as the Dark Brown Flecks. This area is big and has numerous tiny spots that create enormous rings and straps on the fruit's skin. These interpretations focus on regions immediately affecting the fruit and the direct sunlight, mainly in the fruit centre (Filho, 2008) or environmental elements like the precipitation and the intensity of sunshine, together with plant features such as the respiratory capacity and the stage of fruit growth (Reyes *et al.*, 1994).

The increased internal pressure of lactic acid may be related with air (Reyes *et al*, 1994), low temperatures or exceptionally high quantities of heat daily (Duton, 1981). Environmental variables alone do not lead to dull, fruit-type skin problems. As lactic acid near the epidermis is greater (Kaiser *et al.*, 1996) and the soluble fluid is slower, it contains higher ingredients as well, Reyes *et al.*, 1994. All these elements linked to plant material genetic properties are expected to enhance the inner pressure of the lymph cells and therefore to promote the appearance of physiological bruising. (Eloisa *et al.,* 1994) observed that rashes were reduced when they were packaged in white paper bags with immature fruits.

Management

- Thirty days prior to harvest, fruits were packaged in an aluminized polyethylene sheet, minimizing the occurrence of freckles on the fruit.

2. Pulp Flesh Translucency

The so called 'pulp gelling,' 'pulp gelling' and 'pulp gland proliferation' disease causes significant harm to the quality of papaya, as a consequence of high occurrence in papaya growers. Due to the lack of visible signs of the afflicted fruits, problem diagnosis seldom happens in the field. Pulp leaching is frequently only discovered when the fruit is opened, and this presents major marketing difficulties with papaya. In the winter months, the condition is most frequent (Oliveira *et al.*, 2010; Oliveira and Vitoria, 2011). The disturbance comes from the obscuration of the pulp nuance and jelly-like feel or "soaked" (Oliveira and Vitoria, 2011). Symptoms from the endocarp to the exocarp, without any outward process symptoms. The fruit grows denser with the seed cavity building up enormous quantities of water and the fruit starts to sink into the fruit washing tanks. The separation of the body from the skin, in the fruit area, is the most apparent sign of gelling. A specific characteristic of the afflicted fruit. In these circumstances, the papaya pulp can be "peeled" from the skin as if a banana were to be peel.

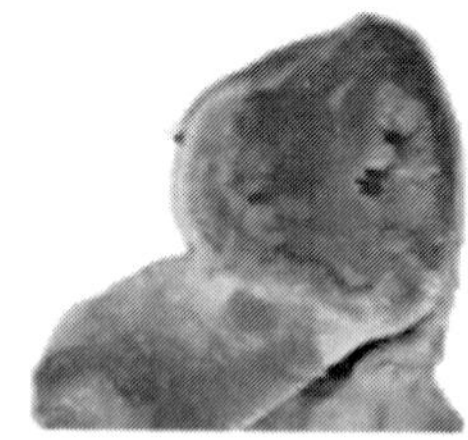

Pulp Flesh Translucency

3. Hard Lumps

Some writers (Paul and Chen 1990; Paul 1995) described the disease known as hard tubes in papaya and its incidence was connected to environmental circumstances, especially high temperatures (Paul, 1995; Oliveira and Vitoria, 2011). This disturbance is characterized by the development of many sections of the mesocarp where the tissue is harder than the surrounding regions. The fruit has no outward signs which differentiate it from unaltered fruits, leading to a reduction in the commercial worth of the fruit during the fruit maturation period, hard lumps. During the summer months (December to March), the problem is difficult to detect in the golden fruit till the remainder of the year. Fruit sensitivity to hard tuber was enhanced throughout the winter months and was substantially linked with fruit fly larvae disinfection and disease control hydrothermal treatment after harvest. Heavy pulp lumps are likely caused by the inactivation of cell wall enzymes owing to high temperature stress which also prevents release of ethylene and alters skin and colour in pulp to papaya. Thermal tolerance can be intentionally generated or develop from natural environmentally stimulating circumstances (climate pre-harvest conditions, in particular temperatures. The production of thermal shock protein (HSP) occurs in papaya upon heat stress and results in thermal tolerance.

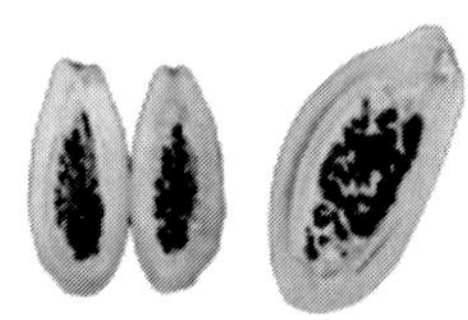

Hard lumps

4. Pulp Softening

The soft papaya tissue generally takes place 6-12 days following the collection of the green and mature goblets (known as discoloration). The tissue varies according to the variety from white, light green to orange, yellow, salmon, or red, and substantial structural changes occur when the tissue becomes soft. Large volumes of cell thin walls that do not endure physical compression.

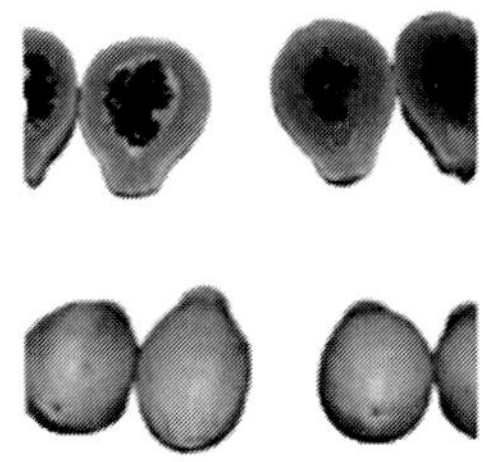

Pulp softening

5. Chilling Injury

This problem arises when fruit is stored under the critical threshold at temperatures or when the storage duration exceeds the recommended practice limit. The temperatures of the fruit vary from one fruit to another and the skin of the fruit is generalized to be more susceptible than pulp (Mitra, 1997). Low-temperature collection of papayas in the fruit and marketing chain is standard practice, particularly if fruit mating is to be delayed away from producing areas for consumption markets.

Chilling injury

Depending on the type, fruit maturing stage and storage, the recommended storage temperature for papaya is between 7 and 13°C. As the fruits mature, they are less prone to cold stress progressively. A ripe, full-color fruit for more than 1 week can be stored for 1 to 3°C. Because papaya fruit may be kept at 7°C for 14 days at the break-even stage, Gomes *et al.* (2016) confirmed that the "golden papaya" held at 10°C for 10 days did not have chills.

Management

- Higher temperature warning treatments can help minimize or eliminate cooling injury during storage.
- Papaya 48.5°C and observed reduced incidence of CI, 4 minutes prior to fruit storage at 5°C.
- Treat papaya with low temperature, decreased cooling stress 6 to 24 hours before transfer to storage at 38°C.
- The tolerance of chilling stress improved before the fruit was stored at 5°C and the preset of Papaya Sunrise to 45°C for 6 hours.
- Alternate temperature for the length of 17 days (two fruit storage periods for the time of 20°C for the 5 days cold storage cycle) did not increase internal damage, but increased skin refreshment due to refreshment.

6. Heat Injury

Heat injuries are caused by changes in the sensitivity of the wood to heat retention during the year and pre- and post-weather preparation conditions. The most common symptoms that occur as a result of heat treatment are outside, that is, darkening or deepening of the skin and accelerating the deposition of "Golden", soft tissue failure in ripening as a result of heat stress. Sometimes the symptoms of a heat injury can be confused with a cold sore injury.

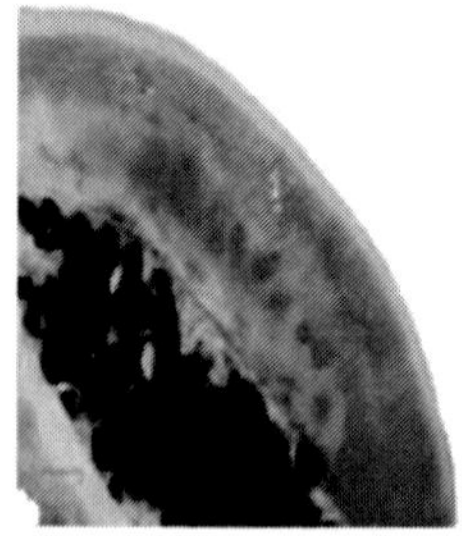
Heat injury

The effect of heat on post-harvest fruit ripening depends, according to Paull and Chen (2000), on (i) of heat tolerance caused by the field, (ii) planting characteristics, (iii) morphology and molecular characteristics of the deposit, (iv) morphological conditions (ripening time), (v) heat transfer rate and equilibrium vi heat dissipation (temperature dissipation) probably only 10-12°C above ambient temperature. The degree of heat injury of wood deposits depends on the temperature and the duration of exposure and how easily wood leaves cool down after heat treatment.

Management

- A tolerance response to fruit undergoing high post-harvest temperatures can naturally be promoted by high pre-harvest temperatures on the orchard.

7. Blossom End Defect

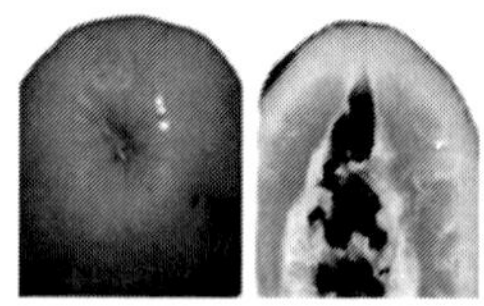
Blossom end defect

The disorder leaves an open channel at the bare tip of the fruit cavity. This situation is caused by the lack of placental expansion in early fruit growth when the carp occasionally does not bind at the end of the ovary to shut the stigmatic channel. Liquido (1990) offers three blister-defeat categories: tiny hole (1 mm diameter), big pinhole (1 mm diameter hole) and navel (1 mm diameter) (similar to the blossom end of a navel orange). The variety "sunrise" with greater carpal fusion due to its elliptic fruit form is more prone to blister defect disturbances. Generally speaking, the fruit that is affected is susceptible to bacterial illnesses in the cavity of the core of the fruit before the fruit that is utilized with the faulty end of the flower is generally matured sooner.

8. Intra-Ovarian Ovaries

Intra-ovarian ovaries

"Intra-ovarian," or "intra-ovarian fruit," is a teratological phenomena that is unusually attractive and documented on papaya. Intra-ovarian fruit does not provide outward signs that make the illness more complicated to detect. Presence in the papaya seed cavity of five tiny secondary fruits. These were connected to the base end of the fruit from the placenta and substituted for the seed. The ovary was surmounted by an incompletely developed immune stige; only one carpel was visible, in each case four of the five secondary foetuses. Each of the four bigger closed fruits included numerous typical seeds, including a normal shape and size embryo. Yellowish and a little plier than typical fruit was the secondary fruit. The epidermis of these fruits was a microscopic analysis consisting of cells of a similar form but smaller than the typical epidermal fruit cells. While this resemblance even pertained to the existence of stomachs, the protecting cells had no chloroplasts. Without a waxy covering and the presence of a thinner epidermis, the cross section of the fruits structure looked like that of a regular fruit.

18

Physiological Disorders of Aonla

1. Internal Fruit Necrosis

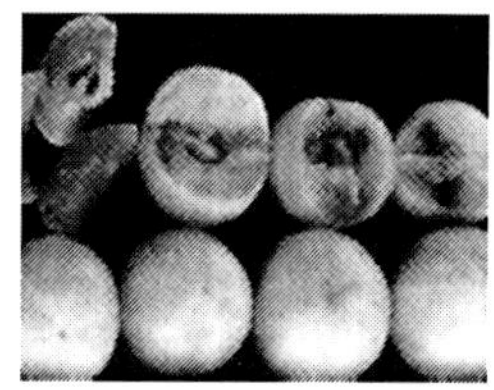

Internal necrosis

Internal necrosis has been observed in Aonla (*Emblica officinalis* Gaertn.) fruits. Francis cultivar, preceded by Banarsi, is highly susceptible. The symptoms begin with browning of the inner majority of the mesocarpic tissue in the 2nd or 3rd week of September at the time of endocarp hardening, which later spreads to the epicarp, resulting in brownish black flesh appearance. Internal fruit necrosis is associated with boron deficiency during the fruit development.

Management

- Infection was not noticed on other Aonla cultivars like Chakiya, NA-6 and NA-7, which need to be encouraged for commercial cultivation.
- Combined zinc sulphate (0.4 percent) + copper sulphate (0.4 percent) and borax (0.4 percent) spray were found to be effective during September October.

2. Fruit drop

Fruit drop

It is a serious problem in Aonla which influences the final yield of crop. There are phases (three waves) of flower and fruit drop, observed in Aonla. In case of the first wave, more than 70% flowers drop within three weeks of flowering, which is normally due to lack of pollination, while during the second wave the drop is that of young fruitlets at the time of dormancy break and in last, *i.e.*, third wave the dropping of fruits is due to embryological and physiological factors and it is spread over the entire period of the fruit development. It starts from latter half of August and continues up to harvest. Fruit drop results due to many factors such as dry spell, imbalance of hormones, improper nutrition, fluctuation in temperature, cultivars and age of tree and number of developing fruits. Also delay in harvesting results in heavy fruit drop particularly in Banarasi and Francis cultivars.

Management

- This disease can be controlled during fruit production by maintaining the required soil moisture level.

3. Off Season Flowering

Off season flowering is also a serious problem in Aonla production. The study conducted by Rai *et al.* (2002) revealed that a complex of situation is responsible for this disorder. An economical loss of 30 to 40% due to off season flowering in Aonla was estimated by them. The main reason of this disorder may be global warming or due to some alteration in the tree physiology due to externally governed internal mechanism which leads to absence of distinct phase of dormancy after fruit harvest. Plant comes in partial flowering with predominance of male flower which do not set fruit, exhaust the plant and disturb the physiology and thereby reduce the crop load in the main season crop. In any case if some fruit set, they do not develop properly and drop subsequently.

Management

- The off-season flowering can be controlled by forcing the plant to enter in dormancy by spraying 10% urea after fruit harvest.
- Restricting soil moisture build up near root zone can lead plant to enter dormancy.

19

Physiological Disorders of Bael and Jackfruit

A. BAEL

1. Fruit cracking

Fruit cracking

Fruit cracking is an important physiological disorder found in some commercial cultivars of Bael (*Aegle marmelos*), which occurs just before the ripening stage. It is correlated with abrupt changes, such as temperature and humidity, in weather conditions. Heavy irrigation or precipitation is also one of the causes of this disease after prolonged drought.

Management

- These can be managed by having excellent irrigation facilities, creating wind breaks around the orchard.
- Spraying borax at full bloom and after fruit set twice @ 0.1 percent.
- Nitrogen and zinc deficiency in Bael orchards are prevalent and can be cured by soil application or foliar spray.

2. Fruit drop

Fruit drop

The large number of flowers and fruits borne by trees are not carried to maturity and a portion of this drop during the course of development. In Bael, fruit drop is a serious problem that occurs due to many variables such as strong winds, soil moisture imbalance, inadequate nutrition and hormonal imbalance.

Management

- This malady can be managed by maintaining the appropriate soil moisture level during fruit development.
- Spraying of growth regulators 2, 4 -D, GA_3 and 2, 4, 5-T at recommended concentrations.

B. JACKFRUIT

1. Irregular shape

The irregularly shaped jackfruit (*Artocarpus heterophyllus* L.) due to Boron deficiency in the fruit is responsible for this. Nutrition plays a critical role and in jackfruit, boron is the secret to enhanced flowering, fruiting and internal and external fruit quality.

Irregular shape

2. Premature fruit drop

Premature fruit drop in jackfruit is often related to unfavourable environmental conditions, irregular watering, improper nutrition and hormonal imbalance. For mature tree watering must be done during dry period from blooming to throughout the growth and development period of fruit. Spray of growth regulators also help to reduce the ill effect of hormonal imbalances.

20

Physiological Disorders of Pomegranate and Sapota

A. POMEGRANATE

1. Fruit cracking

Fruit cracking

Fruit cracking is a serious problem of pomegranate (*Punica granatum* L.) particularly in arid climate. Cracked fruits are sweet but unfit for long distance transportation. Incidence of cracked fruits varies from 10 to 70 % depending upon the prevailing environmental conditions. Longitudinal or radial splitting of rind surface occurs. The longitudinal cracking is more prevalent where the crack originates at the naval or styler end, while radial cracking originates between the ends. The cracked fruits are breeding grounds for insects, harmful bacteria and fungi making fruits unfit for marketing and consumption. The *Mrigbahar* crop is more susceptible to cracking than the crop of other *bahars*. Various factors, including variations in soil moisture regimes, atmosphere, tree nutrition, and cultivars, are responsible for fruit cracking. Young fruits crack due to boron deficiency while fully developed fruits cracks due to differences in soil moisture content, temperature and atmospheric humidity, (Abd El-Rhman, 2010).

Management

- The management strategies include growing of resistant or tolerant cultivars, proper selection of *bahar*, use of Pinolene (5%) as vapour guard, spraying GA3 (120 ppm), liquid paraffin (1%) at 15 days interval, twice during June, application of borax (0.1%) or calcium hydroxide on leaves and fruits after fruit set.
- Adequate calcium and potassium levels as per soil test values need to be maintained.
- Apply adequate and regular irrigation during fruiting season.
- In hot dry period fruits should be covered with butter paper bags.

2. Sun scald

Due to scorching, the surface skin of fruits facing the afternoon sun turns brownish black while the skin underneath is natural.

Sun scald

Management

- To avoid this disorder proper canopy architecture should be developed which prevent the direct exposure of fruits to sunlight.
- Spraying kaolin during the hot summer months is useful in reducing sunscald. First spray of 5% and subsequently 1 or 2 additional sprays with kaolin @ 2.5% at 15 days interval reduce sunscald.
- Bagging the fruits with butter paper covers is useful in minimizing fruit spoilage due to sunscald. White colour bags are more effective in reflecting sunlight and protecting the growing fruits.

3. Aril browning

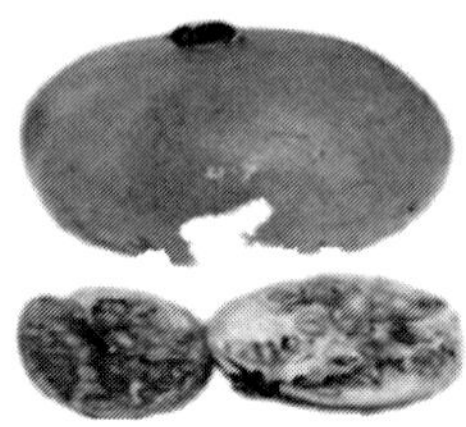

Aril browning

In some widely grown cultivars such as Ganesh and Bhagwa, aril browning or internal breakdown of arils in pomegranate critically affects fruit quality. In certain instances, more than 50 percent of the severity of the disorder in mature ripe fruits may cause significant quality loss. Since the fruits affected by this condition remain free from external signs, it is not possible to isolate them until they are packaged, presenting severe export problems. The aril collapse or browning is characterized by distorted, somewhat flat, dark-blackish and brown arils which have an unpleasant smell when the fruit is opened. This state is followed by desiccation, rustling, and the development of inner arils.

Management

- It can be managed by harvesting at proper maturity period.

B. SAPOTA

1. Fruit drop

The question of low fruit setting and shedding is that of Sapota (*Manilkara zapota* L.) Just around 10-12% of the total fruit set is produced and maintained until maturity. The majority of fruit drops occur immediately after the setting of the fruit.

Management

- By spraying NAA and GA_3 at 25 to 100 ppm during flowering and at 15-day intervals, increases in fruit set and retention are possible.

2. Corky tissue

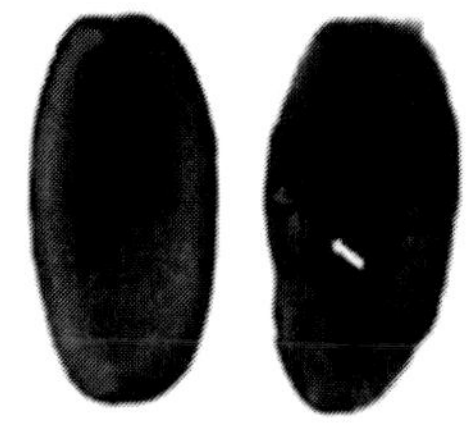

Corky tissue

The rough lump in the pulp, slightly desiccated in nature and acidic to taste, characterizes the corky tissue (CT) of sapota. This disease shows no distinct visible signs and only becomes evident when the fruit is sliced open. Corky skin eruptions are seen under extreme conditions. CT is prevalent in the cultivar 'Cricket Ball' (up to 35%). CT affected fruits have a lower concentration of sugars associated with the accumulation of starch as compared to healthy fruits. Being a climacteric fruit, sapota ripens rapidly under optimal environmental conditions after harvest.

21

Physiological Disorders of Pineapple

1. Multiple Crowns

Multiple Crown

Ordinarily, fruit bears a single crown, but fruit bears more than one in some cases. The top of the fruit will therefore be smooth and wide and the fruit will not be ideal for canning. These fruits have an insipid taste and are corky. It is intended to be an inherited character, often found in the Cayenne group to which the Kew variety belongs.

2. Fruit and Crown Fasciation

Fruit and Crown Fasciation

Fasciated fruits are so deformed that they are entirely useless. Proliferation, in some situations, is so extreme that the fruit is strongly flattened and twisted with countless crowns. Fasciation of fruits and crowns is correlated with elevated plant vigour, which takes longer to bloom. Fasciation can be favoured by high soil fertility, warm weather and calcium or zinc deficiency.

3. Collar of Slips

Collar of slips

The appearance of a large number of slips emerging from the stem near the base of the fruit, or even directly from the fruit itself, characterizes the collar of the slips. At the cost of the fruit, the excessive slip growth results in small, tapered fruits, often with knobs at the base. For such an abnormality, high nitrogen fertilization and high precipitation together with a relatively low temperature are expected to be congenial.

4. Dry Fruit and Bottle Neck

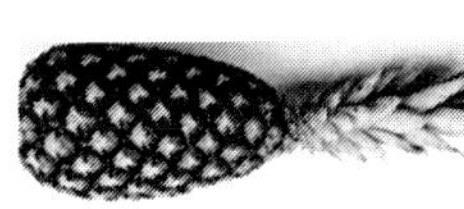

Bottle Neck

The fruit forms of dry fruit and bottle neck are very similar and may be derived from the same parent. The fruit is small in the dry fruit form, flowers are absent, and fruitlets do not grow. Lower fruitlets usually grow in the bottle neck and upper fruitlets do not develop and give the same appearance as dry fruits. Both forms are freely generated by suckers.

5. Sun Scald

This results in plants leaning or dropping to one side, revealing direct sunlight to one side of the fruit. The exposed surface cells get hurt. The surface of the later shell assumes a brownish to black colour and cracks between fruitlets can appear. The fruits that are infected soon rot and become infested with pests. Leaves may be wrapped around the fruits to shield them from sunscalding in ideal climates where leaf development is luxuriant. The other approach is to cover the portion of the fruit exposed to the sun with dry straw or grass or with some other material available locally.

Sun Scald

6. Collar of Slips

The collar of slips is typified by the presence of a large number of slips arising from stem close to the base of the fruit, or even directly from the fruits itself. The excessive slip growth is at the expense of the fruit, resulting in small, tapered fruits, often with knobs at the base. High nitrogen fertilization and high rainfall along with relatively low temperature are supposed to be congenial for such an abnormality.

Collar of Slips

7. Chilling Injury

Chilling injury (CI) occur in pineapple fruits stored below 7°C. External wilting and discoloration of the crown leaves, failure of green-shelled fruit to yellow, browning and dulling of yellow fruit, surface pitting of the fruitlets which eventually coalesce into larger sunken lesions, and postharvest decay. Internal flesh browning or darkening, tissue softening, and off-flavour development. Immature fruits are more susceptible to CI than mature fruit (Full-green colour stage pineapple may suffer CI at 10°C). Although the tissue damage from CI in quarter-yellow colour stage fruits only occurs at temperatures below 7°C, development of the symptoms occurs within several days at ambient temperature following removal from cold storage. Postharvest waxing of the fruit minimizes CI symptom development.

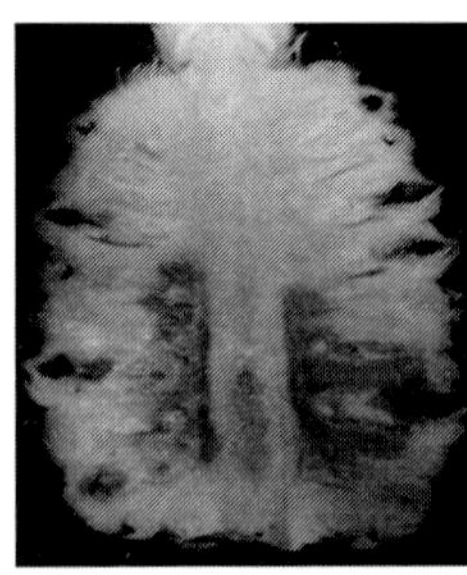

Chilling Injury

8. Internal Browning (Black Heart)

Black Heart

Internal browning is a common postharvest physiological disorder and more pronounced in Smooth Cayenne group varieties that have a low ascorbic acid content. In addition, it is more severe in those fruits exposed to cool night temperatures followed by high day temperatures prior to harvest. Storage of the fruit below 7°C and exogenous application of GA induce this disorder. Symptoms began as small greyish translucent areas at the base of the fruit near the core. These areas eventually darken and in severe cases the internal flesh become firms brown to black in colour. Management of internal darkening at certain extent can be obtained by waxing the fruits.

9. Flesh Translucency

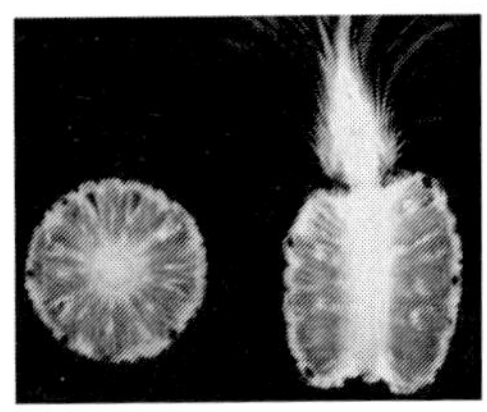
Flesh Translucency

The internal flesh becomes shiny and translucent in appearance. It is more common in less mature harvested fruit with small crowns. Translucency begins before harvest and continues during storage. Waxing can reduce the rate of translucency development after harvest. Translucent fruits are more susceptible to mechanical injury, which leads to intracellular sap leakage and loss of marketable fruit.

MCQ's ON MAJOR PHYSIOLOGICAL DISORDERS OF FRUIT CROPS

1. Mango trees bear a heavy crop in one year (on year) and very little or no crop in the succeeding year (off year) is known as

   a) Spongy Tissue b) Malformation

   c) Alternate Bearing d) Clustering

2. Mango Malformation was first observed in 1891 in

   a) Gujarat b) Bihar

   c) West Bengal d) Maharashtra

3. Black tip disorder of mango was first reported by Woodhouse in

   a) 1906 b) 1907

   c) 1908 d) 1909

4. Alternate Bearing of mango is controlled by

   a) Paclobutrazol b) Deblossoming

   c) Smudging d) All of them

5. Spongy tissue causing fruits loss in mango growing region of India is about

   a) 20% b) 30%

   c) 35% d) 40%

6. More susceptible variety of mango to fruit drop is

   a) Langra b) Dashehari

   c) Fazli d) Amrapali

7. Internal Fruit Necrosis disorder of mango is due to deficiency of

   a) Calcium b) Boron

   c) Zinc d) All of them

8. Leaf scorching disorder of mango is due to deficiency of

   a) K_2SO_4 b) MOP

   c) P_2O_5 d) All of them

9. Higher incidence of fruit cracking in plains of North India was noticed in

   a) Langra b) Dashehari

   c) Fazli d) Amrapali

10. Poor fruiting/ non fruiting/ irregular fruiting are the major problems of

    a) Litchi b) Mango

    c) Banana d) Guava

11. The maximum fruit drop during the first fortnight after fruit set and at harvest the retention varies between

    a) 20-205% b) 33-3032%

    c) 30-396% d) 50-40%

12. Factors influencing the fruit cracking of litchi is

    a) Temperature b) Wind

    c) Maturity d) All of them

13. Lesion browning or pericarp necrosis of litchi is a serious problem in

    a) Litchi b) Mango

    c) Banana d) Guava

14. Browning of the pericarp of litchi occurs at ambient temperatures of

    a) 15-20°C b) 20-30°C

    c) 20-25°C d) 25-30°C

15. Small brown lesions of 2-10 mm in diameter develop in the flesh of the fruit is called as

a) Cork Spot b) Brown Heart

c) Bitter Pit d) Scald

16. Highly susceptible cultivars of Bitter Pit of apple is

a) Golden Delicious b) Marigold

c) Red Delicious d) All of them

17. Less susceptible cultivar of Bitter Pit of apple is

a) Spartan b) Macintosh

c) Gala d) All of them

18. Pre harvest fruit drop of apple can be controlled by application of

a) NAA (15 ppm) b) NAA (5 ppm)

c) NAA (7 ppm) d) NAA (10 ppm)

19. Hen and chicken is physiological disorder of

a) Litchi b) Mango

c) Grape d) Guava

20. Many shot berries surround a bold berry and shot berries is known as

a) Uneven ripening b) Water Berries

c) Hen and chicken d) Shot Berries

21. Presence of green berries in a ripe bunch of coloured grapes is called

a) Uneven ripening b) Water Berries

c) Hen and chicken d) Shot Berries

22. Cluster tip wilting is physiological disorder of

a) Guava b) Grape

c) Mango d) Litchi

23. Choke throat is physiological disorder of

a) Litchi b) Mango

c) Grape d) Banana

24. The chilling of banana occurs when the temperature falls below

a) 12°C b) 13°C

c) 14°C d) 15°C

25. Kottai Vazhai is a serious malady in which variety of banana
    a) Nendran b) Poovan
    c) Basrai d) Robusta
26. Fruit Splitting is major physiological disorder of
    a) Citrus b) Grape
    c) Mango d) Litchi
27. Oleocellosis is more common in crop taken in
    a) Ambe bahar b) Hast bahar
    c) Mrig bahar d) All of them
28. Granulation is major physiological disorder of
    a) Litchi b) Mango
    c) Grape d) Citrus
29. Bronzing is due to the deficiency of
    a) P b) K
    c) Zn d) All of them
30. Bronzing in guava is more severe on
    a) Entisols b) Vertisol
    c) Inceptisols d) None of them
31. Pulp flesh translucency disorder is known as
    a) Pulp flesh gelling b) Pulp gelling
    c) Both a & b d) None of them
32. Formation of highly distinct areas of the mesocarp in which tissue is firmer than surrounding areas is known as
    a) Skin Freckles b) Hard Lumps
    c) Heat Injury d) Chilling Injury
33. Highly susceptible cultivar of Internal necrosis of aonla is
    a) Banarsi b) Chakiya
    c) NA 6 d) Francis
34. Off season flowering is major physiological disorder of
    a) Aonla b) Mango
    c) Grape d) Citrus

35. Spraying borax @ 01% twice at full bloom and after fruit set to control

    a) Browning b) Fruit drop

    c) Fruit cracking d) All of them

36. Irregular shaped jackfruit occurs due to deficiency of

    a) Calcium b) Boron

    c) Zinc d) All of them

37. Premature fruit drop in jackfruit is often related to

    a) Hormonal imbalance b) Irregular watering

    c) Improper nutrition d) All of them

38. Incidence of cracked fruits in pomegranate varies from

    a) 10 to 30 % b) 10 to 40 %

    c) 10 to 70 % d) 10 to 80 %

39. Percentage of the total set fruits develops and retained till maturity in sapota is

    a) 5 to 7 % b) 7 to 10 %

    c) 10 to 12 % d) 12 to 15 %

40. Multiple Crowns is major physiological disorder of

    a) Pineapple b) Pomegranate

    c) Grape d) Banana

Answers

1	c	**9**	b	**17**	d	**25**	b	**33**	d
2	b	**10**	a	**18**	a	**26**	a	**34**	a
3	c	**11**	c	**19**	c	**27**	c	**35**	c
4	d	**12**	d	**20**	c	**28**	d	**36**	b
5	b	**13**	a	**21**	a	**29**	d	**37**	d
6	a	**14**	b	**22**	b	**30**	d	**38**	c
7	b	**15**	c	**23**	d	**31**	c	**39**	c
8	a	**16**	d	**24**	c	**32**	b	**40**	a

Section 4
Major Physiological Disorders of Flowers

22

Physiological Disorders of Rose

1. Blind wood

Blind wood

The normal blooming stem on the greenhouse rose has fully developed sepals, petals and reproductive organs. The inability to build a bloom at the apical end of the stem is a typical phenomenon which in such shoots is termed blind. There are sepals and petals; however, they are not complete or aborted in the reproductive portions. Blind wood is often thin and short, but when it grows at the top of the plant, it can reach considerable length and thickness. This could be caused by inadequate light, insects, rodents, fungal diseases, chemical residues, and other causes.

2. Bull heads or malformed flowers

Malformed flowers

Just partly formed remain the centre petals of the bud and the bud appears smooth. They are prevalent on very vigorous shoots, particularly bottom breaks, and the petals may have a lack of carbohydrates to grow. The cause of bull heading is as yet unknown, but it will also cause malformed flowers to be infested by thrips.

3. Colour fading

Colour fading

For certain yellow varieties, off-coloured flowers pose a problem in that the petals can be green or dirty white instead of pure yellow. The number of off-coloured flowers would be decreased by raising the night temperature by several degrees. The pink or red varicties sometimes producc bluish coloured flowers. The use of organic phosphate and various other kinds of insecticides is also much associated with this.

4. Limp necks

It is a type of disorder in roses caused by water stress in the area just the below the flower head. The affected flower bud bends due to weight of the top.

Insufficient stored energy of flower head may be one of the reasons. Just below the flower "wilts" the area of the stem and will not support the head. This is sometimes due to inadequate water absorption; the flower will be revived by cutting off the lower 1 to 2 inches of stem and putting the cut stem in water at 37°C.

Limp necks

5. Blackening of rose petals

Low temperature and high anthocyanin content are responsible for this. Treatment with GA_3 allows anthocyanin to accumulate in the petals of the Baccara roses. At low temperatures (20°C during the day and 4°C at night), this impact was more pronounced than at higher temperatures (30°C during the day and 20°C at night).

Blackening

6. Bent neck

- It is very common problem of roses and is due to following reason-
- Water deficiency in the neck tissue which is controlled by transpiration rate, the rate of water uptake and the ability of the different organs on the cut flower shoot to complete for water.
- Appearing of plugging materials like Pectin, cellulose and microbes.
- Extreme temperature during shipping or storage.

7. Petal Discoloration

Low night and day temperatures are associated with pigment accumulation and are accompanied by petal blackeningand often stimulated by of ultraviolet radiation through the materials covering the greenhouse. Bluining duration storage is attributed to breakdown of proteins, release of free ammonia and concomitant increase in pH. Bluing of red rose petals is increased by the use of AOA but reduced by the use of STS.

23

Physiological Disorders of Anthurium

1. Sun burn

Leaves in the centres appear bleached and may have brown edges. Shade should be provided to reduce the light level to 1800-2500-foot candles to control this problem.

2. Flower abortion, flower deformation and rosette formation

The flower aborts on the spadix, the spathe growth is stunted, and the flower stem is short. This condition is largely genetically dependent, but can be avoided by minimising the root pressure, encouraging the operation of plants, reducing temperatures, rapid ventilation, cooling and humidification systems. By decreasing irrigation and raising EC of the media, excessive root pressure can be reduced.

3. Cracks

This condition also occurs with temporarily greater relative humidity at the margin of the spathe during the active growth phase. The best possible solution for minimising cracks is lower humidity during the night.

4. Folder Ears

In the early stages, the basal lobe of the flowers will not completely unfold. This condition is based on variety. No solution is known.

5. Sticking

In this case, because the spathe is stuck, the flower does not open. Sticking depends on the variety and happens especially during the rapid growth period. To avoid sticking, lower relative humidity is the best remedy possible. At an early stage, loosening of the flowers by hand may also mitigate this condition.

6. Jamming

Jamming is a powerful variety-dependent feature that occurs more frequently under arid conditions. As the leaves are wrapped too tightly around the flower, the flowers jam into the sheath. Such a condition occurs more often in cultivars with long sheaths. The best remedy is to keep the growing media reasonably moist.

7. Bleached leaves

Sometimes the leaves of Anthurium are found bleaches at the centre and the tips show browning. It is caused by high light intensity or excessive light condition provided to the plant. The shade percentage may be increased to reduce the light level and overcome the problem.

24

Physiological Disorders of Gladioli

1. Fluoride injury

Gladiolus is a fluoride pollution indicator plant, as it expresses the effects of fluorine at very low levels. Symptoms of scorching of the leaves look like drying of the tips of the leaves. A strong application of super phosphate is also associated with fluoride toxicity. By spraying 5 percent lime or Magnesium sulphate, it decreases

Fluoride injury

2. Negative geotropism

This is mostly attributed to auxin's lateral downward motion and its deposition on the lower part of the spike. The spikes should be held vertically in storage as well as during transportation to avoid bending of the tips because of the unequal distribution and vertical transport of auxin.

3. Blindness

Loss of growing point due to unfavourable temperature.

4. Topple

Topple in gladiolus is characterized by breaking of spikes after opening of the florets in the vase. The spikes harvested from calcium-deficient soil have higher incident in topple. Due to Ca deficiency, spray 0.3% $CaCO_3$

5. Tip burn

This physiological condition arises in the atmosphere due to elevated aerial fluoride levels. At the onset of symptoms, a spray of Copper oxychloride (blitox) 50 WP (0.3 percent) should be administered and repeated once more if possible.

25

Physiological Disorders of Carnation and Chrysanthemum

A. CARNATION

1. Splitting of calyx

Splitting

The calyx may be either half or entirely broken. The petals are stripped of their support, which contributes to the petals being bent down. The continuity of the form and arrangement of the flower is thus lost. Splitting, particularly where light and temperature fluctuate is correlated with weather. By maintaining the night temperature at 5 ° F, some reduction in splitting can be achieved. More calyx splits were caused by high plant density per unit area. The number of split calyces was decreased by increasing doses of N, while the rise in potassium rates increased it. Growing of Epson, Palmir, varieties that can withstand calyx splitting.

2. Curly tip

Curly tip

The growing tips that curl and become twisted are affected by this condition. Tips of the young shoots fail to separate and a characteristic curvature result in continued development. The causes of the disease are believed to be poor light and other adverse factors. Water stress and potassium deficiency are suspected causes of carnation flowers dying back from a physiological curly top.

3. Sleepiness

It is disorder in carnation, it mainly occurs by high ethylene synthesis or water stress. Ethylene synthesis occurs much higher when flower is stored for long periods or when they are kept at high temperature.

B. CHRYSANTHEMUM

1. Blindness

Blindness

It happens when the nighttime temperature is too low and when the floral buds grow short. This difficulty is shown by a rosetted form of development. Center petals that do not grow may be due to extreme heat or certain varieties seemingly lack adequate food to open the flower in dark weather. Overwatering, excessive soil fertilizer or insects or diseases that invade the root system are commonly correlated with chlorosis or yellowing of the upper foliage. The mean night temperature was too low when short days began, continued growth of shoots and failure to form flower buds. Sunscald, in very warm weather, is prevalent in flower standards. The petals turn orange, and then they dry out.

2. Leaf Wilting

A lower rate of water uptake caused by bacterial plugging is prevalent in chrysanthemum. Placement of the stems in cold water overcomes the problem.

26

Physiological Disorders of Gerbera, Lilies and Tulip

A. GERBERA

1. Bushiness

An anomaly characterized by numerous leaves, short petioles and small laminae that offer a bushy appearance known as bushiness to some gerbera cultivars. The nodes are not clearly differentiated and no elongation of the internode is seen.

Bushiness

2. Stem break

Sliced gerberas are a common post-harvest disorder. This is primarily caused by imbalances in the water. It may be ethylene-controlled and associated with water stress-related early senescence.

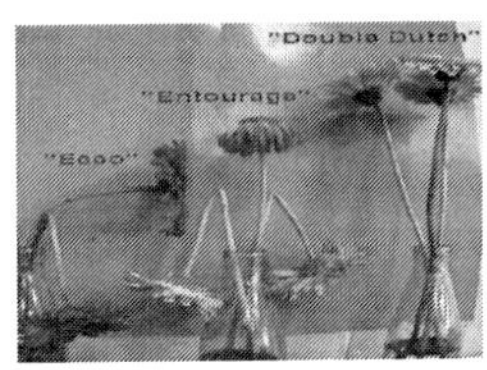

Stem break

3. Yellowing and purple margin

Yellowing and early senescence of leaves are caused by nitrogen deficiency. Phosphorus deficiency with a purple margin causes a light-yellow tint. Increases in nitrogen and phosphorus levels have been observed to encourage sucker production and enhance gerbera flowering.

Yellow & purple margin

B. LILIES

1. Bud drying

It is the common problem of Asiatic hybrids lilies, the apparent desiccation and shriveling of flower bud followed by abscission is a common problem. This problem is solved by STS pulsing.

C. TULIP

Flower bud blasting

A disorder found in tulip that is cause by dry storage or forcing of bulbs. It has been reported a change in the plant's hormonal status, which regulate the distribution in carbohydrates and other organic materials within the plant, is main reason of blasting.

MCQ'S ON MAJOR PHYSIOLOGICAL DISORDERS OF FLOWERS

1. The failure to develop a flower on the apical end of the stem of rose is known as

   a) Colour fading b) Bull heads

   c) Blind wood d) Limp necks

2. Bull heads or malformed flowers is physiological disorder of

   a) Rose b) Carnation

   c) Gerbera d) Anthurium

3. Off coloured flowers of rose is a problem with varieties of

   a) White b) Red

   c) Pink d) Yellow

4. The area of the stem just below the flower "wilts" and will not support the head is called

   a) Colour fading b) Limp necks

   c) Blind wood d) Bull heads

5. Blackening of rose petals is caused by

   a) Low temperature b) Calcium

   c) Zinc d) Boron

6. Carnation varieties tolerant to calyx splitting is

   a) Master b) Epson

   c) Indira d) Star

7. Physiological causes for a curly tip and die back of carnation flowers is

   a) Poor light b) Water stress

   c) Potassium deficiency d) All of them

8. Blindness is a physiological disorder of

   a) Chrysanthemum b) Carnation

   c) Gerbera d) Anthurium

9. An abnormality characterized by numerous leaves, short petioles and small laminae, which give some cultivars of gerbera a bushy appearance known as

   a) Stem break b) Bushiness

   c) Fluoride injury d) Blind wood

10. Stem break of gerbera is mainly caused by

    a) Poor light b) Water stress

    c) Water imbalances d) All of them

11. Yellowing and early senescence of gerbera is due to deficiency of

    a) Nitrogen b) Boron

    c) Calcium d) Potassium

12. Pale yellow colour with purple margin of gerbera is due to deficiency of

    a) Calcium b) Boron

    c) Phosphorus d) Nitrogen

13. Physiological disorder Bent neck related to

    a) Rose b) Carnation

    c) Gerbera d) Anthurium

14. Flower abortion of Anthurium can be prevented by

    a) Rapid ventilation b) Cooling

    c) Humidification d) All of them

15. The best possible remedy to reduce cracks in Anthurium is

    a) Lower humidity b) Ventilation

    c) High humidity d) None of them

16. Folder Ears is a physiological disorder of

    a) Rose b) Carnation

    c) Gerbera d) Anthurium

17. Physiological disorder Sticking related to

    a) Anthurium b) Gerbera

    c) Carnation d) Rose

18. Fluoride toxicity of Gladiolus is associated with heavy application of

    a) Boron b) Nitrogen

    c) Super phosphate d) Phosphorus

19. Fluoride toxicity of Gladiolus is reduced by spraying of
    a) Lime b) Mg sulphate
    c) Both a & b d) Boron
20. Negative geotropism is a physiological disorder of
    a) Gladioli b) Carnation
    c) Gerbera d) Anthurium
21. Loss of growing point due to unfavourable temperature is known as
    a) Topple b) Blindness
    c) Fluoride injury d) Tip burn
22. Topple in gladiolus is due to the deficiency of
    a) Calcium b) Boron
    c) Phosphorus d) Nitrogen
23. Physiological disorder Sleepiness related to
    a) Rose b) Carnation
    c) Gerbera d) Anthurium
24. Bud drying is the common problem of
    a) Chrysanthemum b) Carnation
    c) Gerbera d) Lilies
25. Flower bud blasting is a physiological disorder of
    a) Gladioli b) Carnation
    c) Tulip d) Anthurium

Answers

1	c	**6**	b	**11**	a	**16**	d	**21**	b
2	a	**7**	d	**12**	c	**17**	a	**22**	a
3	d	**8**	a	**13**	a	**18**	c	**23**	b
4	b	**9**	b	**14**	d	**19**	c	**24**	d
5	a	**10**	c	**15**	a	**20**	a	**25**	c

Section 5

Major Physiological Disorders of Plantation Crops

27

Physiological Disorders of Arecanut

1. Band or Hidimundige

Hidimundige

Smaller leaves are produced and finally a rosette is formed. Internodal length reduction and apex-tapering of the stem. Most of the nuts are tiny and deformed, assuming they are produced. Drainage is poor and soil and pan/hard clay pan fertility are low.

Management

- Improved soil management and drainage enhancement.
- Removal of hard pan on the subterranean and foliar micronutrient treatment;
- Soil acidity correction and assimilation of the copper and lime combination.

2. Nut Splitting in Arecanut

Nut Splitting

A ubiquitous physiological problem in nearly every garden it's just physiological. The growth of the pericarp does not match the development of the kernel's internal and so the pericarp and distal end are divided. The nuts are broken down. Contamination of the exposed kernel bacteria and fungi makes nuts unusable after breakage. When 1D 2 to 3D 4 mature, premature yellowing it is commonly seen on immature palms and in parts in individual plantations. The primary reasons include overflow in extremely healthy palms of cell sap, excessive sap flow to the inflorescence and too high supply of nutrients and long-term dryness followed by rapid watering.

Management

- Applying borax to early stages and K_2O to the base is discovered to control to a certain degree the division of the noodles @ 2 g per litre of water (0.20 per cent spray).
- Improved drainage and regular dry-water irrigation.

3. Sun Scorch and Stem Breaking

The beautiful golden dots becoming brown afterwards fissures form at these locations in advanced phases. Moreover, in these sections' saprophytic microorganisms and insects harbour stem breaks later. Palms exposed to southwest sun are harmed due to the harmful influence of solar radiation.

Sun Scorch & Stem Breaking

28

Physiological Disorders of Cashew

1. Little leaf

Little leaf

Little leaf of cashew (*Anacardium occidentale* L.) is caused due to the deficiency of zinc. The leaves remain small in size giving rosette appearance.

Management

- Application of zinc sulphate (0.5%) to the soil or by foliar spray is helpful in preventing this disorder.

2. Yellow leaf spot

Yellow leaf spot

The presence of the yellow leaf spot, an unexplained cashew disease, is associated with low soil pH (4.5-5.0). The presence of yellow leaf spot may be due to molybdenum deficiency. The leaves affected are molybdenum deficient and have disproportionate quantities of manganese.

Management

- Trees sprayed with 0.03 percent ammonium molybdate solution once during the pre-monsoon (June) and after the end of the monsoon (September) were found to be effective in controlling this disorder.
- It can also be controlled by correcting soil pH with lime.

29

Physiological Disorders of Coconut

1. Barren nuts

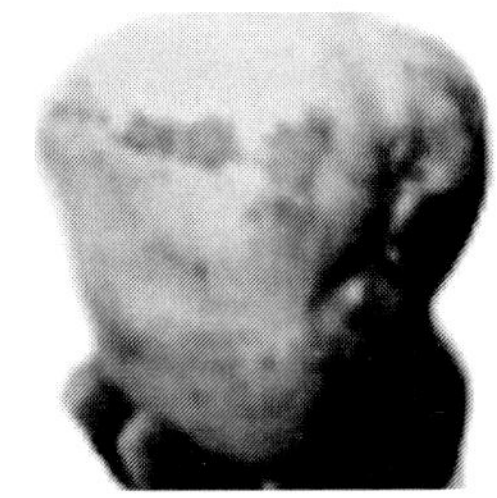

Barren nuts

The presence of nuts in the kernel is known as stubborn nuts or seedless or imperfect nuts and is frequent in coconut (*Cocos nucifera* L.). The nozzles impacted are oblong compared with the typical circular form of the nozzles and less than normal husk is generated. The often-breaking shell over the course of the development is the most typical feature of the barren nut. The major causes attributed to this disorder are the defective fertilization or poor pollination, nutritional deficiency such as K and B in the palm and increased production of palm in the tree.

Management

- The production of barren nuts can be reduced to a greater extent by proper thinning of nuts.
- Adequate application of P and B along with normal recommended dose of fertilizers (1 kg muriate of potash + 200 g of borax + RDF).
- The general practice of applying common salt was also found beneficial in reducing this malady.

2. Button shedding

Button shedding

Button shedding and premature nut fallout usually results in a significant loss in disease yield. This phenomenon of shedding varies from 55-95% depending on prevailing conditions and the cultivation of coconut. The inflorescence fails to fertilize, which does not grow into seeds and eventually sheds. Female flowers or buttons shed after fertilization and some pieces fall prematurely after being set. The fall of immature pieces at a later stage of growth often leads to crop loss. The shedding of potholes is relatively heavy on the dwarf palms rather than the tall ones.

Management

- The occurrence of this malady can be managed by application of proper irrigation by avoiding drought.
- Maintaining good tilth by proper ploughing and better soil aeration.
- Application of potassic fertilizers helps in reducing this disorder.
- Root feeding of coconut with 40 ppm of NAA reduces the button shedding and increases nut yield.
- The effort of application of 2, 4- D @ 30 ppm one month after the opening of the spathe can be made for correcting the ill effect caused of lack of pollination.

30

Physiological Disorders of Date Palm

1. Blacknose

Black nose is used for the unusually shrinking and darker date end. The most susceptible cultivars for this physiological state are Deglet Nour and Hayani. The black nose result is an over exam of the epidermis, in particular many little cross-sectional exams or ruptures near the end of the style of the foetus. There is a substantial shrinkage and darkening in proportion to the quantity of the controls and moist conditions on Khalaal level are related to them. This test is driven by excessive moisture and precipitation; thus, it is necessary to take steps to avoid moisture-enhancing circumstances. In the sensitive period of fruit growth, too much soil humidity and the presence of intercrops and weeds have to be avoided.

Management

- To inhibit the incidence of blacknose testing, bagging the fruits in brown wrapping paper has been found.
- The occurrence of testing and subsequent production of blacknose may also increase over thinning.

2. Crosscuts

Cross or V-cuts in the fruit stalk tissue and a clean split on the cross. It is a deep groove that resembles an artificial cut strand and does not fully develop. In stalks and strawberries, shortcuts originate from anatomical flaws in the inner cavity, which lead to a mechanical breakdown in stalks or seed elongation. And it grows with the older hands. Sayer and Khadrawy varieties are sensitive to this illness and in certain places no longer propagated.

Management

- Crop losses can be averted by the use of non susceptible species or the increase in the number of fruit stalks within sensitive species.

3. Whitenose

In the early stages of Rutab, the dry and prolonged wind causes the fruit to ripen and dry quickly, resulting in a whitish drying at the end of the fruit cup. The affected fruit becomes very dry, hard and has high sugar content.

Management

- In harvested fruits, hydration can correct this condition.

4. Barhee Disorder

Barhee disorders are characterized by irregular bending of the Barhee crown. This phenomenon is extreme and can be bent at an angle of about 90 °. Neither the cause of this disease nor its regulation is understood. However, by providing a strong iron bar on the opposite side of the bend, growers correct this situation; bunches of fruit on the opposite side are attached to this bar to shift the actual weight to the bending side. It seems that the bends will be corrected in 2 to 3 years. To prevent such a deviation, bunch processing is also offered.

5. Black Scald

A small illness of undetermined origin is black scalding, which is different from black nose. It consists of an area that has been blackened and immersed in a certain demarcation line. At the end or side of the fetus, the disease usually occurs and the infected tissue has a bitter taste. The presence of the condition indicates exposure to high temperatures, but the exact cause is certainly not clear.

6. Bastard Offshoot

This is a deformed growth of the vegetative buds of date palms, especially the root roots. Bastard disease is caused by infection of date palm bud ticks. This may also be due to the decline in growth caused by the equilibrium of growth regulators.

7. Leaf Apical Drying

This is not a disease but a physiological response to adult palm transplantation (violation of its root system). After transplantation, all palms with these symptoms recover within two to three years.

8. Fertilization Injury

This type of injury is only seen with palm plants obtained from young tissue culture (the first two years after field planting) and when fertilizers (N, P, K) are applied too close to the palm. The source of fertilizers is not the source, but how the fertilizer is applied near the bar. If the damage is serious, it may cause the young palms to die.

9. Frost Damage

It causes severe metabolic disorders when the temperature drops below 0°C, partial or complete drying of palm leaves, after leaving the cells, the water in the protoplasm freezes. Water penetrates into the intercellular spaces during thawing. The leaves turn brown and dry out. The severity of the damage depends on the intensity and duration of frost.

Frost damage in date palm tree forests has been shown that the fruit of the palm tree is not lost directly, but the freezing and loss of leaves, the palm tree cannot withstand the fruit crop and cannot ripen the following year. About -15°C occurred, and the leaves were completely dry due to frost.

Management

- When the temperature starts to drop low enough (-5°C and below), the most realistic and affordable protection for date growers is to turn on the water and keep the date plant moist.
- Water the date plant only or all irrigates when the temperature drops and retains some heat to provide protection.

10. Lack or Excess of Water

Differences in water supply and water content in the soil significantly affect the growth of date palm. The effect of these water changes can be a reduction in yields or a complete failure in fruit production. Date palm requires an amount of water from 1500 to 2800 mm per year to compensate for the increased evaporation. Prolonged water stress can dramatically reduce growth and yield, and if the drought lasts for many years, the date palm will dry out and die.

On the other hand, if the water level is high and the drainage is insufficient and / or the leaching and transport of soluble salts is incomplete, the salt concentration in the soil and surface water continues to increase at a high rate of evaporation. However, there are salt tolerance thresholds, and if the soluble soil salt is above 6 percent, the date palm does not expand.

MCQ'S ON MAJOR PHYSIOLOGICAL DISORDERS OF PLANTATION CROPS

1. Sun scorch and stem breaking is physiological disorder of
   a) Coconut b) Date palm
   c) Cashew d) Arecanut
2. Little leaf of cashew is caused due to the deficiency of
   a) Nitrogen b) Calcium
   c) Zinc d) Boron
3. Hidimundige is physiological disorder of
   a) Cashew b) Arecanut
   c) Coconut d) Date palm
4. Yellow leaf spot of cashew is caused due to the deficiency of
   a) Molybdenum b) Boron
   c) Calcium d) Nitrogen
5. Nut splitting in arecanut is caused by
   a) Low temperature b) Calcium
   c) Zinc d) Boron
6. The occurrence of nuts without or with imperfectly developed kernel is known as
   a) Barren nuts b) Seedless nuts
   c) imperfect nuts d) All of them
7. The major causes attributed to barren nuts disorder of coconut is
   a) K and B b) P and B
   c) N and B d) None of them
8. Blacknose is a physiological disorder of
   a) Cashew b) Arecanut
   c) Coconut d) Date palm
9. The phenomenon of shedding of buttons of coconut varies from
   a) 50-75% b) 45-95%
   c) 55-95% d) 5-15%
10. Highly susceptible cultivar of Blacknose of Date palm is
    a) Deglet Nour b) Barhee
    c) Khalas d) Thoory

11. Highly susceptible cultivar of Crosscuts of Date palm is
    a) Barhee b) Khalas
    c) Khadrawy d) Yatima
12. Whitenose is a physiological disorder of
    a) Date palm b) Arecanut
    c) Coconut d) Cashew
13. The disease usually appears on the tip or the sides of the fruit is known as
    a) Crosscuts b) Whitenose
    c) Blacknose d) Black scald
14. Barhee disorder is characterized by an unusual bending of the crown of variety
    a) Deglet Nour b) Barhee
    c) Khalas d) Thoory
15. Physiological disorder Bastard off shoot related to
    a) Cashew b) Arecanut
    c) Coconut d) Date palm
16. Leaf apical drying is a physiological disorder of
    a) Coconut b) Date palm
    c) Cashew d) Arecanut
17. Crosscuts incidence increases in palms plant that is
    a) Older b) Younger
    c) Newly d) All of them
18. The severity of frost damage is related to the intensity and duration of
    a) Cold b) Freezing
    c) Frost d) All of them
19. Highly affected crop by Lack or excess of water
    a) Coconut b) Date palm
    c) Cashew d) Arecanut
20. Fertilization injury is a physiological disorder of
    a) Date palm b) Arecanut
    c) Coconut d) Cashew

Answers

1	d	**5**	d	**9**	c	**13**	d	**17**	a
2	c	**6**	d	**10**	a	**14**	b	**18**	c
3	b	**7**	a	**11**	c	**15**	d	**19**	b
4	a	**8**	d	**12**	a	**16**	b	**20**	a

Section 6

Major Physiological Disorders of Spices and Condiments

31

Physiological Disorders of Spices and Condiments

1. Chlorosis

Symptoms of pale or yellow-green leaves (chlorosis) first occur in the older leaves. At the leaf margins reaching toward the midrib or central vein, reddish tints gradually emerge. The leaves are thin. Growth overall is markedly reduced. Deficiency results in rhizome yield loss.

Management

- 1% of urea foliar spray or 2% of DAP twice at weekly intervals.

2. Scorching

Because potassium is very mobile inside the plant, in the event of severe deficiency, symptoms occur only on young leaves. Reduced development, shortened internodes, marginal burning or scorching (brown leaf edges), leaf necrotic (dead) patches, decreased lateral breaks and ease of wilting.

Management

- Application of K @ 3.7 Kg/acre in four splits (basal, 60, 90 and 120 DAP) or K_2SO_4 foliar spray @ 1% at quarterly intervals.

3. Tip burn

Tip burn is a common problem for leaves of many horticultural plants such as coriander, which are associated with significant financial losses. Symptoms include browning of the leaves followed by necrosis at the leaf margin of mature leaves, while water-soaked younger leaves can appear. As a potential cause of tip burn, calcium has long been known, but there is no systematic evidence for its involvement in this syndrome.

4. Edema

Edema (oedema), a common physiological condition, occurs on the undersides of lower or older leaves as bumps, blisters or crystal-like growth. Underneath or probably inside cells, water and solutes build up, causing the epidermal cells to expand and collapse. They do not rupture the cells. It is possible for leaves to turn brownish or tan and become corky. Leaves which are badly damaged will often turn yellow and fall off the vine.

Management

- Usage of well-drained and rising media
- Increasing the intensity of light through spacing plants
- Avoiding over-fertilization, in particular slow growing varieties
- During long periods of low light and cool temperatures, stop watering and growing on the "dry side".
- Placing varieties with similar growth intensity on each irrigation line or section to prevent over-watering again.

5. Browning

Slow plant growth, leaf colour can worsen, thin stems, decreased lateral breaks, loss of lower leaves, browning or purpling of the foliage in some plants.

Management

- It is necessary to add the recommended dose of phosphorus to the soil at the time of sowing or planting.

MCQ's ON MAJOR PHYSIOLOGICAL DISORDERS OF SPICES AND CONDIMENTS

1. Pale or yellow-green leaves first occur in the older leaves is the symptoms of

   a) Chlorosis b) Scorching

   c) Tip burn d) Edema

2. Application of K is recommended for the management of

   a) Chlorosis b) Scorching

   c) Tip burn d) Edema

3. Tip burn is a common problem in

   a) Cardemom b) Clove

   c) Black pepper d) Coriander

4. Blisters or crystal-like growth is found in

   a) Chlorosis b) Scorching

   c) Tip burn d) Edema

5. Recommended dose of phosphorus to the soil at the time of sowing or planting is referred to avoid

   a) Tip burn b) Scorching

   c) Browning d) Edema

Answers

1	2	3	4	5
a	b	d	d	c

References

Abd El-Rhman (2010). Physiological studies on cracking phenomena of Pomegranates. *J. Appl. Sci.,* 6(6): 696-703.

Awneesh (2010). Physiological Disorder Onion-Bolting Agro-pedia, IIT Kanpur, An ICAR NAIP Initiative.

Bhargava, R., Singh, R. S., Pal, G. and Sharma, S. K. (2011). Physiological disorders in fruits in arid region: A review. *Indian J. Arid Hort.,* 6(1-2): 1-10.

Burns, W. and Prayag, S. H. (1920). The book of the mango. Bull deptt., Bombay, p. 103.

Campostrini, E., Pommer, C. V. and Yamanishi, O. K. (2010). Environmental factors causing physiological disorders in papaya plants. *Acta Horticulturae* 851: 453-458.

Chadha, K. L. (2001). Hand book of Horticulture, ICAR, New Delhi.

Chatenet, C., Latché, A., Olmos, E. Ranty, B., Charpenteau, M. and Ranjeva, R. (2000). Spatial-resolved analysis of histological and biochemical alterations induced by water soaking in melon fruit. *Physiologia Plantarum*, 110: 248"255.

Dahsham, D. I. and Habib, S. (1985). Seasonal changes in endogenous auxin- like substances in relation to fruit drop in mango. *Suez Canal University. Is Maileyah,* 2: 769-780.

Davenport, T. L. and Nunez-Eliesia R. (1997). Reproductive physiology (In) The Mango: Botany, Production and uses. pp. 69-146. Litz, R.E. (Ed.), CAB International, Wallingford, UK.

Downton, W. J. S. (1981). Water relations of laticifers in Nerium oleander. Ausralian. *Journal of Plant Physiology* 8: 329-334.

Eloisa, M., Reyes, Q. and Paull, R. E. (1994). Skin freckles on solo papaya fruit. *Scientia Horticulturae*, Amsterdam. 58: 31-39.

Fabi, J. P., Mendes, L. R. B. C., Lajolo, F. M. and Nascimento, J. R. O. (2010). Transcript profiling of papaya fruit reveals differentially expressed genes associated with fruit ripening. *Plant Science* 179: 225-233.

Filho, A. G., Oliveira, J. G., Pio Viana, A. and Pereira. (2008). Skin freckles and yield components of papaya Tainung 01: Effect of irrigation depths and soil coverings. *Ciencia e Agrotecnologia*, 4: 1161–1167.

Gunjate, R. T., Tare, S. J., Rangwala, A. D. and Limaye, V. P. (1979). Effect of preharvest and postharvest calcium treatment on calcium content and occurrence of spongy tissue in alphanso mango fruits. *Indian J. Hort.,* 36(2): 140-144.

Gupta, S.N. (2019). Instant horticulture. Jain Brothers. ISBN: 978-81-8360-115-3.

Jawanda, J. S., Singh, Raghbir and Pal, R. N. (1974). Effect of growth regulators on floral bud drop, fruit character and quality of "Thompson Seedless" grape (*Vitis vinifera* L.). *Vitis,* 13: 215-221.

Joshi, G. D. and Roy, S. K. (1985). Spongy tissue in mango, A physiological disorder. *Indian Hort.,* 29(4): 21-22.

Kaiser, C., Allan, P., White, B. J. and Dehrmann, F. M. (1996). Some morphological and physiological aspects of freckle on papaya (*Carica papaya* L.) fruit. *Journal of South African Society Horticulture Science* 6: 37–40.

Katrodia, J. S., Rane, D. A. and Salunkhe, D. K. (1988). Biochemical nature of spongy tissue in alphanso fruits. *Acta Horticulturae*231: 835-839.

Kumar, R. and Kumar, K. K. (2007). Managing physiological disorders in litchi. *Indian Horti.,* 52(1): 22-24.

Ladania, M. S. (2008). Phsiological disorders and their management. (In) Citrus fruit: Biology, Technology and evaluation Ladania M.S. cr citrus. pp. 451- 463.

Mishra, S. K., Naurial, J. P. and Awasthi, R. P. (1973). Effect of growth regulator on fruit drop in litchi. *Punjab Hort. J.,* 13: 122-126.

Muthukumar, P., Selvakumar, R. (2013). Glaustas Horticulture, New Vishal Publications, Delhi, India, ISBN : 8183990711.

Nirvan, R. S. (1953). Bunchy top of young mango seedlings. *Sci. Cult.,* 18: 335-336.

Oliveira, J. G., Bressan-Smith, R. E., Campostrini, E., Da Cinha, M., Costa, E. S., Netto, A. T., Coutinho, K. S., Silva, M. G. and Vitória, A. P. (2010). Papaya pulp gelling: is it premature ripening or problems of water accumulation in the apoplast? *Revista Brasileira de Fruticultura,* 32(4): 961-969.

Pandey, R. M. and Sharma, Y. K. (1979). Leaf scorch- A nutritional disorder in mango. *Indian J. Hort.,* 36(1): 110-113.

Paull, R. E. (1995). Preharvest factors and the heat sensitivity of field-grown ripening papaya fruit. *Postharvest Biology and Technology,* 6: 167-175.

Paull, R. E. and Chen, N. J. (1990). Heat shock response in field grown, ripening papaya fruit. *Journal of American Society of Horticulture Science* 115: 623-631.

Prasad, A., Singh, H. and Shukla, T. N. (1965). Present status of mango malformation disease. *Indian J. Hort.,* 22: 254-265.

Queitsch, C., Hong, S. W., Vierling, E. and Lindquist, S. (2000). Heat shock protein 101 plays a crucial role in themotolerance in Arabidopsis. *The Plant Cell* 12: 479-492.

Rai, M. Vishal Nath, Singh, H. S., Dwivedi, R. and Gangopadhaya, K. K. (2002). Overcoming off season flowering in aonla. *Indian Horticulture* 47(3): 12-13.

Ravindra, V. and Shivshankar, S. (2006). Spongy tissue in alphanso mango. II. A key evidence for the causative role of seed. *Curr. Sci.,* 91(12): 1712-1714.

Reyes, Q., Eloisa, M. and Paull, R. E. 1994. Skin freckles on solo papaya fruit. *Science Horticulture* 58: 31-39.

Schripsema, J., Vianna, M. D., Rodrigues, P. A. B., Oliveira, J. G. and Franco, R. W. A. (2010). Metabolomic investigation of fruit flesh gelling of papaya fruit (*Carica papaya* L. Golden) by nuclear magnetic resonance and principle component analysis. *Acta Horticulturae*, 851: 505-511.

Sharma, R. R. (2005a). Fruit drop causes and control. In: Fruit Production-Problems and Solution, Indian Agricultural Research Institute, New Delhi, pp. 55-71

Sharma, R. R. and Saxena S. K. (2004). Rootstocks influence granulation in Kinnow mandarin (*Citrus nobilis* and *C. deliciosa*). *Scientia Hort.,* 101: 235–242.

Sidhu, A. S. (2008). Storage potential of some onion and garlic cultivars: A comparative evaluation. *Indian J Hort.* 65: 347-48.

Singh, B. N. and Chakravarti, S. C. (1935). Observation on a disease of mango at Banaras. *Sci. Cult.,* 1: 294–295.

Singh, H. P. (1986). Cracking of fruits a problem in litchi growing. *Chona Hort.,* 3(2): 8-9.

Singh, R. (2001). 65-Year research on citrus granulation. *Ind. J. Hort.,* 58(1/2): 112–144.

Singh, R. N. (1990). Mango, Indian Council of Agricultural Research, New Delhi, p. 134.

T. K. Bose (2002). Vegetable Crops, Volume-1, Naya Prakash, ISBN 8185421749.

Zhang, C., Huang, H. and Kuang, Y. (1995). A study of the cause of the mango black tip disorder. *Scientia Horticulturae*64: 49-54.

Zong, R. F., Shao, P. F., Hu, X. Q. and Dai, L. Y. (1979). Preliminary studies on fluctuations in the components of the juice sac and rind in citrus fruit granulation. *Scientia. Agri. Sincia,* 3: 60-64.

Colour Plates

Chapter 3: Physiological Disorders of Solanceaus Vegetables

Blossom end rot

Fruit cracking

Catface

Sunscald

Blotchy ripening

Physiological leaf roll

Herbicide injury

Goldspeck

Russeting

Edema

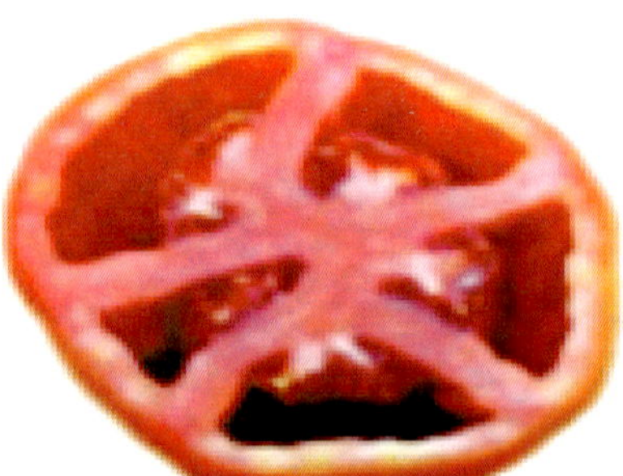

Puffiness

Zebra Stripe

Zippering

Blossom end rot

Sunscald

Blossom end rot

Blackheart

Tuber Cracks

Pink pigmentation

Greening

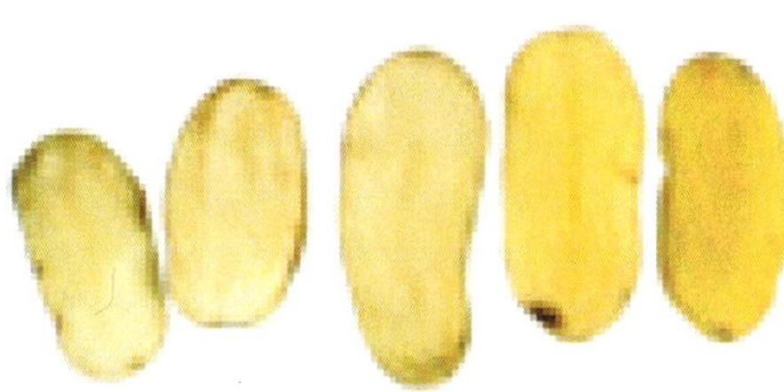

White Flesh

Brown heart

Enlarged lenticels

Secondary growth

Internal Brown Spot

Chapter 4: Physiological Disorders of Cole Crops

Browning

Whiptail

Buttoning

Blindness

Bursting

Tip burn

Leafiness

Riceyness

Fuzziness

Hollow stem

Chapter 5: Groundwater, Wells and Pumps

Poor seed germination

Bolting

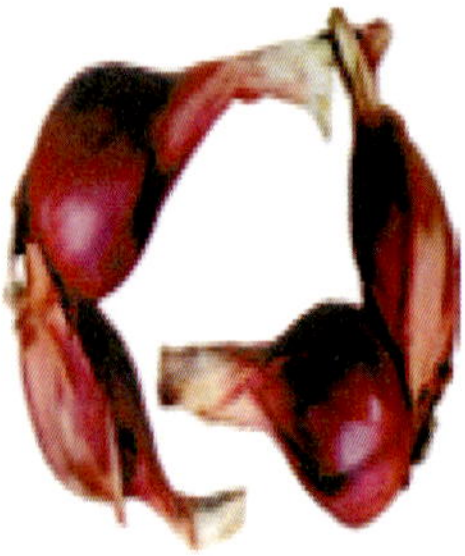

Splitting

Sprouting

Greening

Sprout inhibitor injury

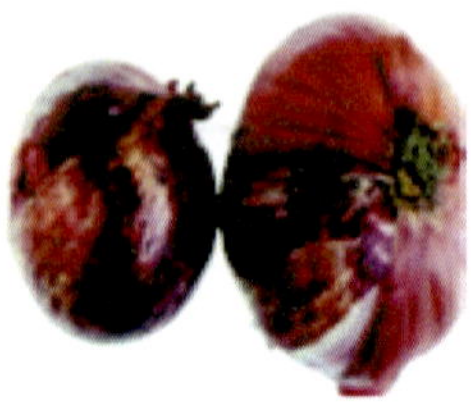

Sunscald

Translucent scale

Pelting rain injury

Tip-burn

Ozone injury

Herbicide injury

Chapter 6: Physiological Disorders of Root Crops

Forking

Pithiness

Bolting

Root splitting

Cavity spot

Internal browning

Brown heart

Zoning

Black heart

Cracked Stem

Pencil strip

Chapter 7: Physiological Disorders of Leafy Vegetables

Tip burn

Russet spotting

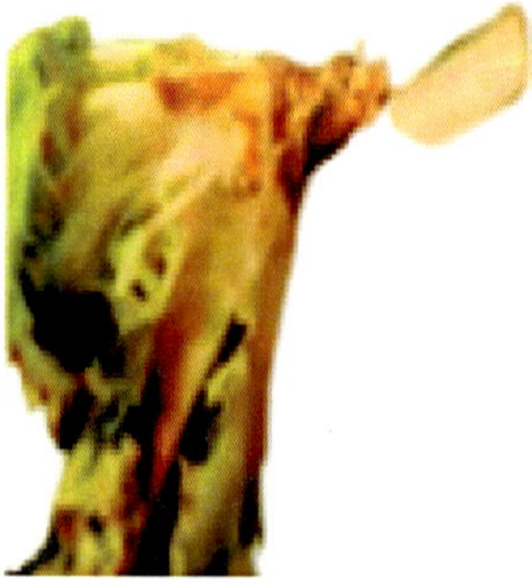

Brown Rib

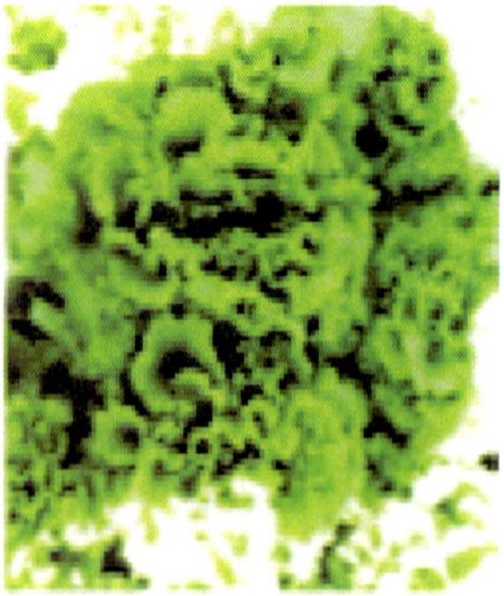

Ribbiness

Rib discoloration

Chapter 8: Physiological Disorders of Cucurbitaceous Vegetables

Blossom end rot

Drought stress

Flood damage

Poor pollination

Oedema

Pillow

Leaf Silvering

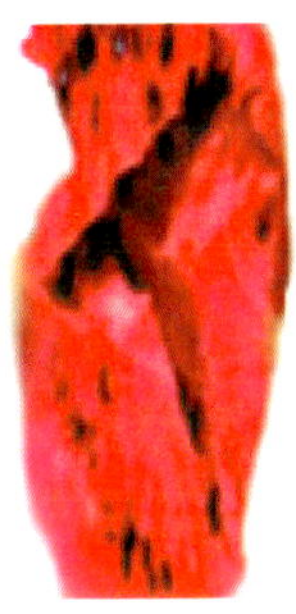
Hollow heart

Rind necrosis

Delay in ripening

Stem splitting

Chilling injury

Measles

Sunscald

Unfruitfulness

Chapter 10: Physiological Disorders of Mango

Malformation

Black tip

Spongy Tissue

Fruit Drop

Fruit Necrosis

Clustering

Cracking

Chapter 11: Physiological Disorders of Litchi

Poor Fruiting

Flower drop

Cracking

Sunburn

Underdeveloped Fruits

Black Spot

Chapter 12: Physiological Disorders of Apple

Bitter pit

Brown heart

Cork Spot

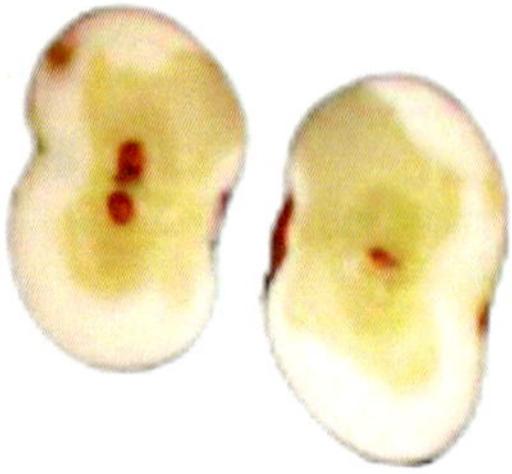

Water Core

Sun burn

Russeting

Fruit drop

Chapter 13: Physiological Disorders of Grapes

Hen and chicken

Uneven ripening

Water Berries

Cluster tip wilting

Shot berries

Pink berry

Bud and flower drop

Poor cane maturity

Chapter 14: Physiological Disorders of Bananas

Choke throat

Chilling injury

Kottaivazhai

Finger drops

Cracking

Chapter 15: Physiological Disorders of Citrus

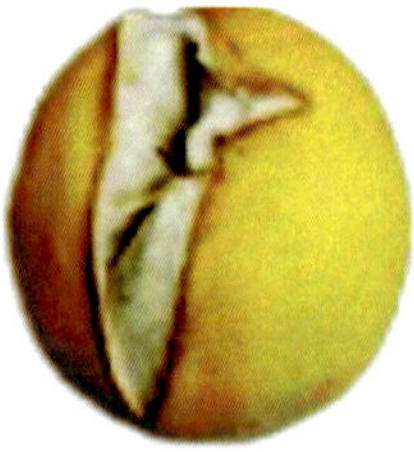
Fruit Splitting

Fruit drop

Oleocellosis

Granulation

Citrus decline

Puffing

Chapter 17: Physiological Disorders of Papaya

Skin freckles

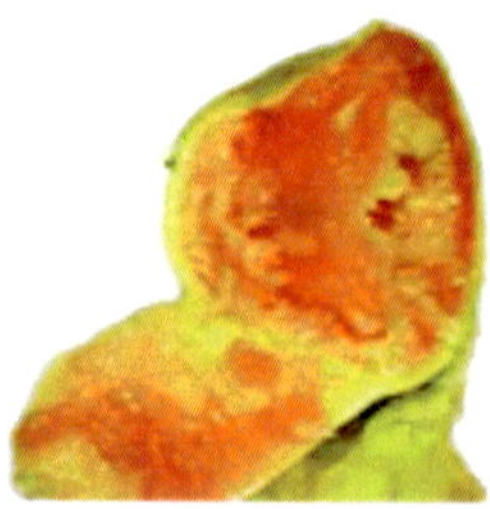
Pulp Flesh Translucency

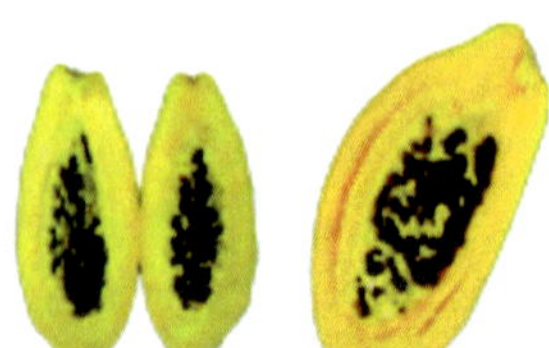
Hard lumps

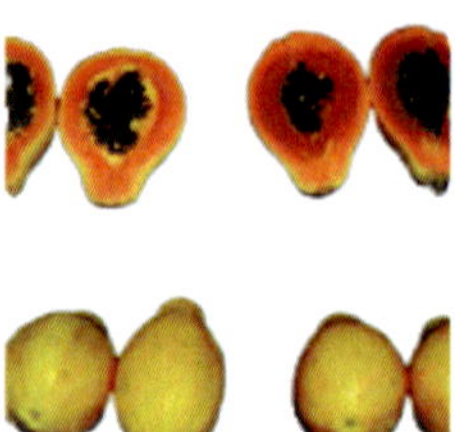
Pulp softening

Chilling injury

Heat injury

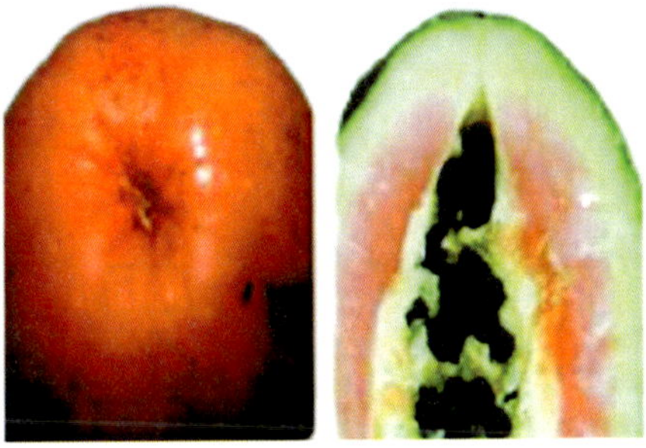

Blossom end defect

Intra-ovarian ovaries

Chapter 18: Physiological Disorders of Aonla

Internal necrosis

Fruit drop

Chapter 19: Physiological Disorders of Bael and Jackfruit

Fruit cracking

Fruit drop

Irregular shape

Chapter 20: Physiological Disorders of Pomegranate and Sapota

Fruit cracking

Sun scald

Aril browning

Corky tissue

Chapter 21: Physiological Disorders of Pineapple

Multiple Crown

Fruit and Crown Fasciation

Collar of slips

Bottle Neck

Collar of Slips

Sun Scald

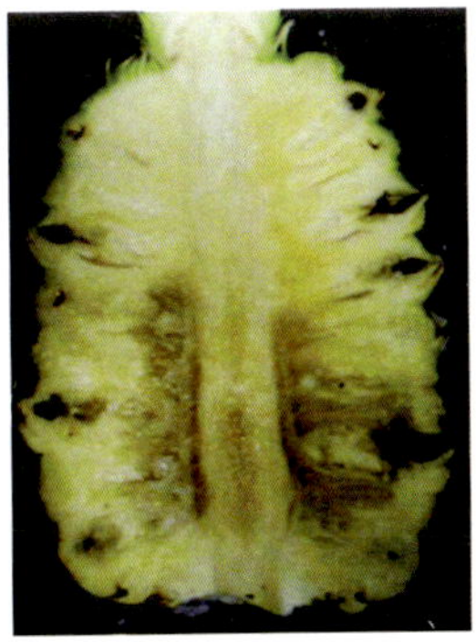

Chilling Injury

Black Heart

Flesh Translucency

Chapter 22: Physiological Disorders of Rose

Blind wood

Malformed flowers

Colour fading

Limp necks

Blackening

Chapter 25: Physiological Disorders of Carnation and Chrysanthemum

Splitting

Curly tip

Blindness

Chapter 26: Physiological Disorders of Gerbera, Lilies and Tulip

Bushiness

Stem break

Yellow & purple margin

Chapter 27: Physiological Disorders of Arecanut

Hidimundige

Nut Splitting

Sun Scorch & Stem Breaking

Chapter 28: Physiological Disorders of Cashew

Little leaf

Yellow leaf spot

Chapter 29: Physiological Disorders of Coconut

Barren nuts

Button shedding

Appendices*

*Start on next page

Appendix-I

Major physiological disorders in horticultural crops

S. No.	Name	Cause	Control measure/Remarks
A. Fruits:			
1.	Mango		
(i)	Black tip	Due to smoke of brick kilns+deficiency of Boron	Borax spray (0-6%)
(ii)	Malformation	Due to low temperature Bombay green-highly susceptible	Resistant varieties-Bahaduran Illaichi, Mangeera, Aliff
(iii)	Spongy tissue	Due to heat convection	Resistant variety: Ratna, Arka, Arka Puneet, Arka Anmolm major problem in Alphonso
(iv)	Alternate bearing	Alphonso is highly susceptible	Cultar (PBZ) application
(v)	Jhumka (clustering)	Low temperature in February – March+ Improper pollination & fertilization.	NAA @ 200-300 ppm
(vi)	Leaf scorch	Chloride toxicity, Excess Potassium	Potassium sulphate 5 %
(vii)	Internal necrosis	Boron deficiency	Highly susceptible-Dashehari Free-Neelam, Langra
(viii)	Jelly seed	-	Highly susceptible-variety Tomy atkins
(ix)	Taper tip	-	Deshehari is susceptible
(x)	Girdle necrosis	-	
(xi)	Softnose	Excess of Calcium and Nitrogen deficiency	-
2.	Banana		
(i)	Neer vazhai (water banana)	-	Nendran-susceptible
(ii)	Kotta vazhai (seed banana)	-	Poovan-susceptible.

(iii)	Improper bunch filling	Deficiency of Potassium	-
3.	Citrus		
(i)	Granulation	High temperature & Relative humidity during ripening	Spray of lime
(ii)	Leaf mottling/frenching	Zinc deficiency	-
(iii)	Exanthema/ Ammoniation of dieback	Copper deficiency	-
(iv)	Yellow leaf of citrus	Molybdenum deficiency	-
4.	Guava		
(i)	Bronzing	Zinc deficiency	L-49 is more tolerant than Allahabad Safeda
5.	Grape		
(i)	Hen & Chicken disease	Boron deficiency	0.2% boric acid
(ii)	Millerandage in Grape	Boron deficiency	-
(iii)	Berry or Blossom drop in Grape	Improper pollination & fertilization	1.0 per cent calcium nitrate
(iv)	Coulure & shot berries	Boron deficiency, Improper pollination	Application of GA
(v)	Calyx end rot	Calcium deficiency	-
(vi)	Pink berry formation and its clone	High temperature	Common in Thomson seedless
(vii)	Uneven ripening	Gulabi, Bangalore blue, susceptible	Application of Ethephon (250 PPM)
6.	Litchi		
(i)	Little leaf+Leaf bronzing	Zinc deficiency	-
(ii)	Fruit cracking	Excessive water +high temperature	2,4-D and NAA at 20 ppm
7.	Aonla		
(i)	Fruit Necrosis	Boron deficiency	Francis-highly susceptible Chakaiya-Resistant variety
8.	Pomegranate		
(i)	Fruit cracking	Boron deficiency+moisture imbalance more in *Mrige bahar*	Tolerant varieties: Khandari, Alandi, Bedana, Bosec

(ii)	Internal break down	More in Ambe Bahar	
9.	Persimmon		
(i)	Calyx end rot	Calcium deficiency	-
10.	Avocado		
(i)	Dry neck	-	-
11.	Apple		
(i)	Water core	Boron deficiency	Avoid delayed harvests
(ii)	Bitter pit +cracking	Calcium deficiency	Calcium sprays prior to harvest
(iii)	Wither tip	Copper deficiency	-
(iv)	Rossette leaves	Zinc deficiency	Adequate irrigation, manuring
(v)	Interveinal chlorosis	Magnesium deficiency	-
(vi)	Endoxerosis (June dorp)	Water deficiency	-
(vii)	Early drop	Improper pollination and fertilization.	-
(viii)	Scald	Storage at high temperature	-
(ix)	Internal browning	-	Susceptible variety: Yellow Newton
(x)	Storage breakdown	Susceptible variety: Golden Delicious, yellow newton	
(xi)	Jonathan spot	Water deficiency	-
12.	Pear		
(i)	Black end	Water deficiency	-
(ii)	Pear scald	Prolong handing of fruits in storage.	Susceptible: Conference
13.	Apricot		
(i)	Tip burning	Due to high temperature	-
(ii)	Apolexy	-	-
14.	Mangosteen		
(i)	Gamboge	Heavy rainfall	Yellow exudation of gum on fruits
15.	Pineapple		
(i)	Multiple crown (Fasciation)	Genetical/Nutritional factor	Susceptible variety Kew
(ii)	Sun scald	Exposure of fruit to sun rays	-
16.	Strawberry		
(i)	Albinism	Lack of fruit colour during ripening	-

17.	Sapota		
(i)	Cock's comb	-	-
18.	Custard		
(i)	Stone fruits	High relative humidity, low temperature	-
19.	Peach		
(i)	Woolliness	-	-
B. Plantation Crops:			
20.	Cocoa		
(i)	Cherelle wilt	-	-
21.	Coffee		
(i)	Black bean/ Black Jolloo/ Normal Jolloo	-	
22.	Rubber		
(i)	Tapping Panel dryness (TPD) or Brown Blast	Excessive harvesting of latex	Tapping rest
23.	Tea		
(i)	Tea yellow disease	Sulphur deficiency	-
24.	Coconut		
(i)	Crown chocking	Boron deficiency	-
(ii)	Little leaf	Zinc deficiency	-
25.	Arecanut		
(i)	Band disease	Improper drainage	-
C. Vegetables:			
26	Tomato		
(i)	Blossom End Rot (BER)	• Lack of calcium • Wide fluctuations in soil moisture levels • Excessive nitrogen fertilization	• Spray of 0.5 % calcium chloride • Increase in the level of phosphate • Liming • Avoid over-fertilizing with nitrogen
(ii)	Fruit cracking	Dry period is followed by heavy rains and high temperatures	• Uniform supply of moisture • Plant crack-resistant varieties, such as 'Jetstar.' • Soil application of Borax @ 15-20 kg per hectare

			• Spraying of Borax 0.25 %
(iii)	Catfaced fruit	• Cloudy and cool weather at blooming time	• Maintenance of sufficient soil moisture. • Recommended cultural practices should be adopted.
(iv)	Sunscald	• High fruit pericarp temperature 40ºC • In bright sunlight, surface temperature may be more then 10ºc highest then the air temperature.	• Utilize cultivars with enough foliage to cover the fruit • Crop are planted at higher densities are less susceptible • Cultivation of indeterminate /semi-determinate varieties without staking
(v)	Blotchy ripening	• K deficiency	• Use of balanced fertilizer dose • Adjust planting date to achieve favorable light intensity
(vi)	Physiological leaf roll	• Irregular water supply	• Planting determinate cultivars. • Maintaining uniform, adequate soil moisture. • Avoiding severe pruning. • Maintaining temperatures below 95°F
(vii)	Unfruitfulness	• Both high and low temperature	• Growing high temperature tolerant varieties HS-102, Punjab Kesri, Punjab Chuhara, Hot Set. • Growing low temperature tolerant varieties Pusa Sheetal, Cold Set. • Application of PCPA @ 50 ppm at full bloom stage or 2,4-D @ 2 ppm
(viii)	Puffiness	• Non-fertilization of ovules • Embryo abortion after normal fertilization • High temperature and high soil moisture	• Maintenance of normal temperature. • Spraying of Borax 10-15 ppm at the peak flowering time.
(ix)	Gold Fleck or Gold Speck	• Excess calcium • Temperature increases	• Increasing the electrical conductivity of the nutrient solution • Cultivars resistant to BER
(x)	Blossom Drop	• Extremes in temperature and dry conditions • Night temperatures are below 55 F or above 75 F.	• Water the plants deeply
27.	Brinjal		
(i)	Poor fruit set	• High temperature	• Spray 2 ppm 2,4, D

		• Unproductive flower	• Spray 60 ppm NNA or 50 ppm PCPA
28.	Potato		
(i)	Blackheart	• Over irrigation • Poor ventilation • High temperature	• Avoid harvesting in extreme heat (around 30°C) or excessively low temperatures (around 0°C). • Maintain a moderate temperature during cold storage. • Maintain adequate ventilation during transport
(ii)	Tuber cracks	• Irregular irrigation	• Not planting susceptible cultivars. • Regular and appropriate irrigation • Moderate supply of nitrogen
(iii)	Greening	• Exposed to light	• Ensure a spacing of 90 cm between rows • Keep the seed tubers in the centre of the ridge. • Irrigate in order to prevent ridge destruction
(iv)	Brown heart and hollow heart	• Excess fertilization • Planting is too late or early	• Do not plant too early • Use moderate fertilization. • Ensure a good tuber initiation
(v)	Enlarged lenticels	• Tubers spend long time in wet soil	• Irrigate regularly but not excessively. • Efficient drainage. • Dry the tubers thoroughly at harvest time.
29.	Cauliflower		
(i)	Ricyness	Fluctuation in temperature, High humidity, Excess Nitrogen	• Select proper variety and transplanting at right time
(ii)	Fuzziness	Cultivation in abnormal time	• Sowing of good quality of seed in right times
(iii)	Blindness	Frost	• Very low temperature should be avoided
(iv)	Leafyness	High temperature	• Select right variety for the particular growing condition
(v)	Buttoning	Nitrogen deficiency, sowing early variety in late season	• Adequate supply of nitrogen and Right selection of variety
(vi)	Hollo stem	Excessive Nitrogen	• Borax 1.1 to 3.4 kg per hectare.
(vii)	Whiptail	Molybdenum deficiency	• Sodium Molybdate (10-15 kg/ha)
(viii)	Browning or Brown rot or Red rot	Boron deficiency	• 10- 15 kg borax/ha

30.	Onion & Garlic		
(i)	Bolting	High temperature and excess N	• Moderate temperature at bulbing and supply recommended dose of nitrogen
(ii)	Bulb sprouting	Excess Nitrogen + soil moisture	• Spray Chlorpropham or Isopropyl N-(3-Chlorophenyl) Carbamate (CIPC) @ 2% at 75 days after transplanting.
(iii)	Splitting	Delays Harvesting	• Plant onion at recommended spacing, avoid over fertilization and uneven irrigation
(iv)	Greening	Exposed to sunlight	• Do not cure bulbs for extended periods in the field
31.	Beet root and Radish		
(i)	Hollow root	High temperature	
(ii)	Internal browning	Heat stress	Selection of varieties according to their adaptability and seasons
(iii)	Brown heart or crown heart or heart rot	Boron deficiency	10-15 kg of borax per hectare
(iv)	Akashin	Boron deficiency	1-2 ppm of boron
(v)	Pungency	High temperature and water stress	Proper water management.
32.	Carrot		
(i)	Splitting	Boron deficiency, Excessive Nitrogen, Change in soil moisture	Harvest at right maturity stage
(ii)	Cavity spot	Calcium deficiency, Excessive Nitrogen, Change in soil moisture	Increase Ca level
(iii)	Bitterness	Excessive ethylene	
(iv)	Forking	Hard soil pan	Varieties with vigorously elongated root system should be selected
(v)	Pithiness	-	Harvesting at appropriate time
33.	Celery		
(i)	Black heart	High nitrogen, potassium, and sodium levels	Applications of soluble calcium
(ii)	Cracked stem	Calcium deficiency	Application of calcium
(iii)	Pencil strip	Excess of Phosphorus	Optimum application of phosphatic fertilizers
34.	Lettuce		
(i)	Tip burn	Ca deficiency + Unfavourable climate	Harvesting at proper time
(ii)	Rosette spotting	Ethylene injury	Not storing in same room with ethylene producing products

(iii)	Brown Rib	High temperatures and humidity	Grow resistant cultivars
(iv)	Puffy Heads/ Ribbiness	High temperature, rain and high humidity	Use of adapted cultivars
(v)	Rib discoloration	High temperature	Irrigate during hot weather
35.	Watermelon		
(i)	Blossom end rot	Heat stress + moisture stress + Ca deficiency	Provide recommended calcium and proper irrigation.
(ii)	Hollow heart	Low boron	Supply adequate boron
(iii)	Rind Necrosis	Drought stress	Avoid drought stress in melon
36.	Cucumber		
(i)	Pillow	Calcium deficiency	Supply proper dose of calcium
(ii)	Light Belly Color	Cool, moist soil	Avoid excessive nitrogen
37.	French Bean		
(i)	Blossom drop	High temperature	Maintain normal temperature.
(ii)	Hypocotyl cracking (necrosis)	Calcium deficiency	Supply adequate calcium
D. Flowers:			
38.	Rose		
(i)	Bullhead	-	
(ii)	Bent neck (limp neck)	Storage disorder	
(iii)	Sleepiness	Ethylene injury during transportation	
39.	Carnation		
(i)	Calyx splitting	If night temperature falls below 10°C	Common in double carnation
(ii)	Slabside	Cool periods	
40.	Chrysanthemum		
(i)	Premature budding	-	
(ii)	Quilling of florets	-	
41.	Gladiolus		
(i)	Topple	Calcium deficiency	
(ii)	Budrot	Calcium deficiency	
(iii)	Blindness	Unfavourable temperature	
(iv)	Negative geotropism	Due to uneven distribution of auxin	
42.	Lilium		
(v)	Leaf scorch	Manganese Aluminium deficiency	
(vi)	Bud Blast	-	

Appendix-II

Major plant Growth Regulators used in horticultural crops

Active Ingredient	Formulation	Concentrate	Product Name	Crops
Auxins				
2-(1-naphthyl) acetic acid	SL	45 gram/l	Planofix	Pineapple, Apple, Pear
4-indol-3-ylbutyric acid	DP	1 gram/kg	Seradix B No 1	Ornamental crops
4-indol-3-ylbutyric acid	DP	3 gram/kg	Seradix B No 2	Ornamental crops
4-indol-3-ylbutyric acid	SL	8 gram/kg	Seradix B No 3	Ornamental crops
Gibberellins				
Gibberellins	SL	32 gram/l	ProGibb 4%	Mango, Citrus, Grape, Grape-seedless, Pear, Potato, Hops
Cytokinin's				
6-benzyl adenine/gibberellins	SL	19/19 gram/l	Promalin	Plum, Apple, Ornamental crops, Flowers
Ethylene				
Ethephon	SL	480 g/l	Ethrel	Cherry, Peach, Pineapple, Plum, Prune and Sugarcane Citrus, Cotton, Apple, Grape, Maize,
Growth Retardants				
Paclobutrazol	SC	250 g/l	Cultar	Litchi, Mango, Peach, Plum
Daminozide	SP	850 g/kg	B-Nine SP	Flowers, Ornamentals
Glyphosate-isopropylamine	SL	360 g/l	Glyphosate 360 Acid	Sugarcane and Grasses (chemical mowing)
Glyphosate-isopropylamine	SL	360 g/l	Mamba 360 SL	Sugarcane Grasses (chemical mowing)
Glyphosate-isopropylamine	SL	360 g/l	Roundup	Grasses (chemical mowing) Sugarcane
Glyphosate-isopropylamine	SL	360 g/l	Roundup Ultra	Sugarcane Grasses (chemical mowing)
Growth Inhibitors				
Mepiquat chloride	SL	50 g/l	Pix	Cotton
Chlormequat chloride	SL	750 g/l	CeCeCe 750	Pear,
Chlormequat chloride/ethephon	SL	300/150 g/l	Uprite	Wheat
Defoliants				
Thidiazuron/diuron	SC	120/60 g/l	Dropp Ultra	Cotton
Growth Stimulators				
Brassionosteroid	SL	0.1%	Double	Agril. & Horticultural crops

SL-Liquid Suspension; WP-Wettable Powder; EC-Emulsifiable Concentrate; DP-Powder Dust

Fruit Ripening Plant Growth Regulators

Crop	Chemical	Dose (ppm)	Reaction
Mango	Ethephon	1000	Accelerated fruit maturation and surface colour
Citrus	Ethrel	1000	Improvement Induce seven days yellow colour.
Banana	Ethrel	1000	Two days of accelerated maturation
Papaya	Ethephon + NaOH	2000	24 hours of maturation
Sapota	Ethephon + NaOH	500	Mature in two days

Flowering and Fruit Set Plant Growth Regulators

Crop	Chemical	Dose (ppm)	Spraying time	Response
Coconut	2,4-D	30	One month after spathe opening, micro sprayer within a week of the last bud opening	The percentage of fruit setting grew to 32.5% compared to 23% in control
Banana	2,4-D	25	Flowering spray	Prevents seediness in Poovan cultivar.
Mandarin Orange	2,4-D or NAA	20, 100	Calyptra falling dip cluster	Fruit setting increases
Grape	GA3	25	In the stage 35-40.	Fruit setting increases in grape cultivar Thompson Seedless.
Pineapple	Planofix 10 ppm +2% urea 0.04% sodium carbonate +20 ppm ethrel	50 ml/plant 200 to 300	Spray at fruiting	Uniform flowering induced to increase fruit size
Snake gourd, Bitter gourd, Bottle gourd	Ethrel	100	After sowing 4 times 10-15 days at weekly intervals	Yield increases
Ridge gourd, Pumpkin	Ethrel	250	After sowing 4 times 10-15 days at weekly intervals	Yield increases
Sweet potato	Ethrel	250	15 days after planting at fortnight interval	Tuber yield increases

Appendix-III

Average Annual Temperatures & R.H. in Indian states

States/ UT's	High (°C)	Low (°C)	R.H. (%)
Madhya Pradesh	32	19	50
Chandigarh	30	17	-
Uttarakhand	28	15	=
Uttar Pradesh	32	18	56
Punjab	30	17	65
Uttarakhand	19	8	66
Delhi	31	19	54
Bihar	31	20	-
Himachal Pradesh	17	10	-
Jammu and Kashmir	20	7	70
Tripura	31	20	-
Orissa (Odisha)	33	22	70
Assam	29	19	79
Nagaland	22	15	-
West Bengal	32	22	71
Chhattisgarh	33	21	-
Jharkhand	29	18	-
Meghalaya	21	12	81
Tamil Nadu	32	21	70
Telangana	32	21	56
Lakshadweep	31	25	77
Karnataka	30	19	65
Andaman & Nicobar	30	22	79
Pondicherry	33	24	-
Kerala	31	23	78
Andhra Pradesh	33	24	72
Gujarat	34	21	55
Rajasthan	32	19	44
Goa	32	23	76
Maharashtra	32	18	53

Appendix-IV

Essential Nutrients for Plants

Basic elements	Mineral Elements Macro-nutrients Primary			Macro-nutrients Secondary	
C	N	Ca	B	Mn	Al
H	P	Mg	Cu	Mo	Co
O	K	S	Fe	CI	Na
	Zn	Ni			
	Si				
	V				

Appendix-V

Soil Nutrient Index: Macro and Micro-Nutrients State-wise Distribution

S. No	STATES / UT's	CYCLE I (2015-2017)			CYCLE II (2017-2019)		
		Low	Medium	High	Low	Medium	High
1	A & N Islands	N, P, K, OC, S	B, Cu, Fe, Mn, Zn	N, P, K, OC, S	B, Cu, Fe, Mn, Zn		
2	Andhra Pradesh	N, B, Fe, Zn	K, OC, Cu, Mn, S	P	N, Zn	OC, B, Cu, Fe, Mn, S	P, K
3	ArunachalPradesh	P, B, Zn	K, Cu, Fe, Mn, S	N, OC	P, K, B, S, Zn	Cu, Fe, Mn	N, OC
4	Assam	P, K, B	N, OC, Cu, Fe, Mn, S, Zn	P, K, B	N, OC, Cu, Fe, Mn, S, Zn		
5	Bihar	N, B, Cu, S, Fe, Mn, Zn	P, K, OC	N, B, Fe	P, K, OC, Cu, Mn, S, Zn		
6	Chhattisgarh	N, B, Zn	P, K, OC, S, Cu, Fe, Mn,	N, S, Zn	P, K, OC, B, Cu, Fe, Mn		
7	D & N Nagar Haveli	N, P	OC, B, Cu, S, Fe, Mn, Zn	K	N, P	OC, B, Cu, S, Fe, Mn, Zn	K
8	Daman And Diu	N, P, B	K, OC, Cu, S, Fe, Mn, Zn				
9	Delhi	N, P, OC, Fe, Mn	K, B, Cu, S, Zn	P	N, OC, B, Cu, Fe, Mn, S, Zn		K
10	Goa	P, B	N, K, Cu, Fe, Mn, S, Zn	OC	P, B, S	N, K, Cu, Fe, Mn, Zn	OC
11	Gujarat	N, B, Fe, S,Zn	P, K, OC, Cu, Mn	N, B	P, OC, Cu, Fe, Mn, S, Zn		K
12	Haryana	N, P, OC	K, B, Cu, Fe, Mn, S, Zn	N, P, OC, B, Fe, Mn	K, Cu, S, Zn		
13	Himachal Pradesh	N	P, K, B, Cu, Fe, Mn, S, Zn	OC	N	P, K, B, Cu, Fe, Mn, S, Zn	OC
14	Jammu And Kashmir	P, B, Fe, Mn, S, Zn	N, K, OC, Cu	P, Mn, S, Zn	N, K, B, Cu, Fe	OC	
15	Jharkhand	N, P, B, S, Zn	K, OC, Cu, Fe, Mn	N, P	K, OC, B, Cu, Fe, Mn, S, Zn		

16	Karnataka	N, B, Fe, S, Zn	P, K, OC, Cu, Mn	N, B, Fe, S, Zn	P, K, OC, Cu, Mn		
17	Kerala	N, B, S	P, K, Cu, Fe, Mn, Zn	OC	N, B, S	P, K, Cu, Fe, Mn, Zn	OC
18	Madhya Pradesh	N, P, Zn	K, OC, B, Cu, Fe, Mn, S	N, P, Zn	K, OC, B, Cu, Fe, Mn, S		
19	Maharashtra	N, OC, B, Fe, S, Zn	P, Cu, Mn,	K,	N, B, Fe, S, Zn	P, OC, Cu, Mn	K
20	Manipur	N, P, K, Fe, Mn	OC, B, Cu, S, Zn	N, P, B, S, Zn	K, Cu, Fe, Mn	OC	
21	Meghalaya	N, P, K, B, Fe, S	Cu, Mn, Zn	OC	N, P, K, Mn, S, Zn	B, Cu, Fe,	OC
22	Mizoram	P, OC, B	N, K, Cu, Fe, Mn, S, Zn	P, OC, Cu, Mn	N, K, B, Fe, S, Zn		
23	Nagaland	P, Cu, Zn	K, B, Fe, Mn, S	N, OC	P, Zn	K, B, Cu, Fe, Mn, S	N, OC
24	Odisha	N, P, OC, B, Cu, Fe, Mn, S, Zn	K	N, P, OC, B, Cu, Fe, Mn, S, Zn	K		
25	Puducherry	N, P, OC, B	K, Cu, Fe, Mn, S, Zn	N, P, B	K, Cu, Fe, Mn, S, Zn	OC	
26	Punjab	N, P, OC, B, Mn	K, Cu, Fe, S, Zn	N, P, OC, B, Mn	K, Cu, Fe, S, Zn		
27	Rajasthan	N, OC, B, Fe, S, Zn	P, K, Cu, Mn	N, OC, B, Fe, Zn	P, K, Cu, Mn, S		
28	Sikkim	B, Mn	N, P, K, Cu, Fe, S, Zn	OC	N, B	P, K, Cu, Fe, Mn, S, Zn	OC
29	Tamil Nadu	N, OC, B, Fe, S	P, K, Cu, Mn, Zn	N, OC, B, Fe, S	P, K, Cu, Mn, Zn		
30	Telangana	N, P, OC, B, Cu, Fe, Mn, S, Zn	K	N, OC, Fe, Zn	P, K, B, Cu, Mn, S		
31	Tripura	P, K, S	N, OC, B, Cu, Fe, Mn, Zn	N, K	P, OC, B, Cu, Fe, Mn, S, Zn		
32	Uttar Pradesh	N, P, OC, B, S	K, Cu, Fe, Mn, Zn	N, P, OC, B, S	K, Cu, Fe, Mn, Zn		
33	Uttarakhand	N, B	P, K, OC, Cu, Fe, Mn, S, Zn	N, B	P, K, OC, Cu, Fe, Mn, S, Zn		
34	West Bengal	N, B, S, Zn	K, OC, Cu, Fe, Mn	P	S	N, K, B, Cu, Fe, Mn, Zn	P, OC